Cooperative Decision-Making Modeling for Manned and Unmanned Aerial Vehicles

Focusing on "human-in-the-loop" and "human-on-the-loop" mechanisms, cognitive intelligent interaction, and the practical modeling, implementation, and application of limited human intervention in modern aerial systems, this book explores cooperative decision-making technologies for manned and unmanned aerial vehicles.

It systematically discusses two primary approaches to cooperative decision-making mechanisms: limited human intervention and cognitive intelligent interaction. The authors introduce modeling methods for applying limited intervention in typical mission scenarios such as obstacle avoidance, threat mitigation, and attack decision-making. They also explain how to assess human workload and cognitive load in manned and unmanned operations. Additionally, it outlines interactive cognitive models that support situational awareness, threat evaluation, task allocation, route planning, and decision simulation. These insights address the growing need for effective human-machine collaboration in complex operational environments driven by rapid advancements in information science, control science, cognitive science, and artificial intelligence.

This title will appeal to researchers, engineers, and professionals specializing in command-and-control systems and intelligent decision-making systems. It will also serve as an essential reference for students and educators in information-related disciplines.

Jun Chen is a professor at the School of Electronics and Information, Northwestern Polytechnical University, China. His research focuses on modeling and practical applications in intelligent decision-making for complex autonomous systems, command-and-control systems, and human-machine interaction, with a strong record of academic publications in these fields.

Xinyu Zhang is an associate professor at the School of Electronics and Information, Northwestern Polytechnical University, China. Her research interests span AI, data science, data mining, human-machine interaction, and deep learning applications, in which she has published a series of papers.

Yan Tong is currently pursuing a doctoral degree at the School of Electronics and Information Technology, Northwestern Polytechnical University, China. His research focuses on modeling collaborative decision-making between manned and unmanned systems using knowledge- and data-driven fuzzy cognitive maps.

Cooperative Decision-Making Modeling for Manned and Unmanned Aerial Vehicles

Jun Chen, Xinyu Zhang, and Yan Tong

CRC Press
Taylor & Francis Group
Boca Raton London New York

CRC Press is an imprint of the
Taylor & Francis Group, an **informa** business

SCIENCE PRESS

Designed cover image: Shutterstock (Asset ID: 1164487207)

First edition published 2027
by CRC Press
2385 NW Executive Center Drive, Suite 320, Boca Raton FL 33431

and by CRC Press
4 Park Square, Milton Park, Abingdon, Oxon, OX14 4RN

CRC Press is an imprint of Taylor & Francis Group, LLC

ISBN: 978-1-041-34179-6 (hbk)
ISBN: 978-1-041-34240-3 (pbk)
ISBN: 978-1-003-79171-3 (ebk)

DOI: 10.1201/9781003791713

Typeset in Minion
by Apex CoVantage, LLC

Contents

Preface

WITH THE ADVANCEMENT OF unmanned aerial vehicle (UAV) technology, the incremental adoption of UAV assets, and their expanding operational applications, the integration of UAVs with traditional manned aircraft to leverage their complementary strengths has emerged as a significant research focus both domestically and internationally. Cooperative decision-making represents a key area of interest in the control science and engineering domains, serving as a key technology for the command and control of manned-unmanned teaming operations. In the collaborative control process between manned and unmanned platforms, the human operator should avoid direct intervention in the UAV's low-level control systems, instead directing cognitive resources toward higher-level planning, decision-making, and supervisory tasks. This shift in control focus from lower-level skill-based execution to higher-level cognitive engagement is termed the "supervisory control" paradigm. Given the heterogeneous characteristics of manned-unmanned cooperation, effective collaboration must adhere to specific coordination mechanisms that effectively harness the operator's cognitive capabilities, that is, enhancing the comprehension, analysis, judgment, and prediction of the battlefield situation, while fully leveraging the UAV intelligent decision-making system's substantial potential in sensing, computation, and data storage.

With rapid interdisciplinary advances in cognitive science, information science, control science, and artificial intelligence, research on human-autonomous intelligent system collaboration has become increasingly urgent. In response, this book addresses future trends in manned-unmanned cooperative decision-making by proposing two collaborative mechanisms, including the limited-intervention and the cognitive-intelligent interaction approaches. This book further details the approaches and implementation methods for each and is structured into eight chapters. Chapter 1 introduces the research background, significance, and current developments domestically and internationally. Chapter 2 presents the limited-intervention cooperative decision-making mechanism. Chapter 3 describes the limited-intervention cooperative decision-making model. Chapter 4 introduces the cognitive-intelligent interactive cooperative decision-making mechanism. Chapter 5 covers the workload and cognitive load evaluation models for manned aircraft operators. Chapter 6 addresses human-machine collaborative situation awareness assessment and threat assessment models based on cognitive-intelligent interaction. Chapter 7 discusses task allocation and path-planning models under the same cognitive-intelligent interaction framework. Chapter 8 presents a simulation system for manned-unmanned cooperative decision-making based on cognitive-intelligent interaction.

This book summarizes the authors' research output in this field in recent years, addressing key issues of current interest and reflecting the latest research advances and trends. The technologies, methods, and models presented are closely aligned with actual equipment capabilities and operational scenarios, fully capturing the defining characteristics of "human-in-the-loop" and "human-on-the-loop" in manned-unmanned teaming.

During the preparation and proofreading stage of this book, Professor Xiaoguang Gao provided consistent support and guidance. Several graduate students, including Xinwei Zhang, Qilin Zhang, and Teng Lei, participated in the research work related to this publication. We would also like to acknowledge PhD candidates Xinru Duan and Jiaqi Zhang for their contributions in translating and proofreading the book.

The book translation was primarily conducted by human translators. Generative AI tools (specifically, Gemini 3, DeepSeek) were utilized exclusively for the purposes of proofreading and terminology reference. These tools were not used to generate substantial academic content or original text.

The publication of this book was supported by the National Natural Science Foundation of China (Youth Program No. 61305133), the Aeronautical Science Foundation (No. 2016ZC53020), and the 2021 Northwestern Polytechnical University High-Quality Academic Monograph Cultivation Project.

Due to the inherent limitations of the authors' expertise, certain shortcomings in the content may be unavoidable. The authors sincerely welcome constructive feedback and corrections from readers.

CHAPTER 1

Introduction

THIS CHAPTER PRIMARILY ELABORATES on the research background, significance, and state of the art of collaborative decision-making for manned and unmanned aerial vehicles (UAVs), domestically and internationally, as well as the common modeling and analytical methods involved.

1.1 RESEARCH BACKGROUND AND SIGNIFICANCE

1.1.1 Research Background

In recent years, UAVs have undergone rapid development in various fields. As a new combat force, their role has expanded beyond traditional domains of information support and assurance, like reconnaissance, surveillance, and hydro-meteorological operations, to include firepower strikes with more offensive and lethal capabilities. UAVs are expected to undertake more critical missions in future warfare [1]. Consequently, the advancement of UAV technology and the proliferation of related equipment have prompted significant global research into how to effectively integrate this new force into combat systems and maximize its advantages in future warfare [2–4]. However, given the increasingly complex and intense battlefield environment, along with the multiplicity and complexity of tactical missions, it is challenging for UAVs to independently accomplish missions in high-intensity confrontations. Therefore, to meet operational requirements for UAVs in future warfare and account for the distinct combat characteristics of manned and unmanned platforms, manned-unmanned teaming enables seamless information and resource sharing. This approach supports information-driven, networked,

DOI: 10.1201/9781003791713-1

intelligent, and systemic combat operations across the entire combat ecosystem [5]. This approach maximizes the strengths of both asset types and holds significant military value for enhancing overall system combat effectiveness [6].

Manned-unmanned cooperation involves the coordinated operation of one or more manned aircraft with one or more UAVs to execute various combat missions. Within such hybrid formations, the unmanned elements may range from high-performance, multi-role large UAVs to lower-performance, single-function small UAVs. Information exchange between manned and unmanned platforms is conducted via tactical data links, jointly constituting an integrated team. Beyond inheriting certain characteristics of traditional combat units, manned-unmanned formation substantially extends the forms. Consequently, it faces operational challenges such as establishing system-level cooperation modes and collaborative rules, as well as technical challenges such as developing the advanced technologies needed for effective system connectivity. The foundation for effective manned-unmanned teaming rests on three key prerequisites, including a manned platform with simple and effective UAV control capabilities, UAVs that possess a sufficient degree of autonomy, and an efficient, reliable data-linking system [7].

1.1.2 Research Significance

1.1.2.1 Necessity of Manned-Unmanned Cooperative Combat

1.1.2.1.1 Adapting to the Rapid Development of UAV Technology and Enhancing Its Combat Capabilities under Highly Contested Conditions [8]

Since the beginning of the 21st century, UAVs represented by the U.S. "Global Hawk" and "Predator" have demonstrated outstanding performance in successive counterterrorism wars and other military operations, playing an increasingly critical role. In October 2001, during the war in Afghanistan, the U.S. military carried out a live-ammunition attack using "Predator" UAVs armed with Hellfire missiles, marking a major shift in UAV missions from traditional reconnaissance to direct fire strikes. However, these successful operations were conducted under conditions where the U.S. military had full air superiority and faced little aerial opposition. Despite continuous advances and improvements in UAV technologies, UAVs still exhibit certain inherent limitations, constrained by the level of autonomous capability. At present, UAVs mainly perform missions independently and have not yet achieved cooperative combat capability under highly contested conditions. When facing adversaries with strong countermeasures, the employment of UAVs poses substantial operational risks.

Under such a context, the concept of "swarm warfare" has emerged, that is, multiple UAVs coordinate in a manner analogous to a bee swarm, achieving collective combat capabilities that substantially exceed those of individual UAVs operating in isolation. Yet, from a practical point of view, the technology for cooperation among UAVs remains immature. There are significant challenges that persist in cooperative situational awareness, information exchange, and intelligent decision-making, presenting a substantial gap for practical deployment and widespread adoption. In contrast, cooperative combat utilizing a hybrid formation of manned aircraft and UAVs presents lower technical barriers and a more feasible path to realization than fully autonomous swarms. Consequently, nations worldwide today are prioritizing the development of manned-unmanned teaming technologies.

1.1.2.1.2 Achieving Complementary Advantages While Improving Combat Efficacy The operational characteristics and employment modes of manned aircraft and UAVs differ significantly. Manned aircraft possess strong comprehensive judgment capabilities, whereas UAVs have limited intelligent decision-making ability. Manned aircraft exhibit strong rapid response capabilities, while UAVs have lower levels of judgment and identification, relying on manual control from a ground control station. Manned aircraft incur high costs and risks if lost, while UAVs are low-cost, attrition-tolerant alternatives with no risk of human casualties. Manned aircraft have relatively short flight durations and limited sustained combat capability, whereas UAVs possess long endurance and strong sustained combat capability, to name a few.

Manned-unmanned hybrid formations enable role complementarity, allowing each platform to fully leverage its inherent strengths. On one hand, UAVs can function as tactical vanguards, capitalizing on inherent advantages such as low cost and reduced observability. On the other hand, human operators contribute higher-order cognitive intelligence and integrated judgment to compensate for UAVs' operational limitations in high-threat environments. For instance, in an air-to-ground strike scenario, the manned aircraft operates remotely from beyond the enemy's air defense engagement zone, providing command and control. It is responsible for processing complex battlefield information, rapidly performing tasks like target identification and engagement decisions, while simultaneously providing necessary aerial cover for the UAVs. Under the command of a manned aircraft, UAVs can leverage their low observability to strike inland targets, thereby enhancing both overall combat efficacy and battlefield survivability. The loss of a UAV is a more tactically acceptable risk than the loss of a manned aircraft and its crew. Furthermore, the use of inter-aircraft data links for command and control alleviates the burden on the theater's satellite communication infrastructure, enhancing operational resilience.

1.1.2.2 Advantages of Manned-Unmanned Cooperative Operations

1.1.2.2.1 Further Strengthen Network-Centric Warfare and Enhance System-Level Operational Efficacy Network-centric warfare focuses on integrating three core functions into a unified information network, including intelligence collection and fusion, command and control with information exchange, and firepower strike, thereby converting formerly dispersed sensing, command, and weapon systems into a single, efficient, networked system. The modern aerial battlefield has evolved from a platform-centric paradigm, which prioritized individual platform performance and firepower, toward a network-centric model emphasizing information superiority, system integration, and cooperative effects through multiplatform joint operations. A fundamental operational advantage of UAVs stems from their inherent node characteristics. As unmanned airborne platforms, their flight operations and mission execution depend entirely on the integration of onboard systems, external information sources, and rear-echelon command and control. When UAVs form hybrid formations with manned aircraft for large-scale cooperative operations, the establishment of clear tactical relationships serves to enrich and solidify the connections between combat units and other network nodes. This integration enables the full realization of systematic warfare capabilities while preserving baseline platform performance. Furthermore, this synergy will expand the operational modalities of network-centric warfare, ultimately driving a shift from a platform-centric, through a network-centric, and finally toward an information-centric warfare paradigm.

1.1.2.2.2 Spawn New Operational Modes and Promote a Transformation in Aerial Combat Paradigms The transformative impact of UAVs on operational modes is evident in the combat deployment of the U.S. military's "Global Hawk" and "Predator". The Global Hawk's exceptional endurance and operational range have fundamentally transformed traditional reconnaissance methodologies, which previously depended on the pre-conflict positioning of manned aircraft and complex multi-aircraft relay rotations. The Predator system fundamentally reconfigured the traditional find-locate-strike sequence by integrating reconnaissance and strike capabilities into a single platform, thereby pioneering the modern integrated reconnaissance-strike operational model. When manned and unmanned aircraft jointly form combat units, they can not only execute diverse missions like air-to-air combat, air-to-ground operations, reconnaissance and surveillance, and electronic warfare but also expand formation combat paradigms. The focus of the formation shifts from maintaining a strict external configuration toward leveraging the integrated advantages of its constituents. Within a manned-unmanned hybrid formation, UAVs can extend the traditional formation combat model across the entire observe-orient-decide-act (OODA) cycle. They can surge forward to attract enemy fire, serve as communication relays, provide high-altitude auxiliary target detection, maintain level flight for weapon guidance, or conduct stealthy assaults for direct strike missions. Traditional operational methods, such as suppressive fire penetration and aerial cover, will become obsolete. This change will not only alter the methods of aerial combat but, more significantly, will drive a fundamental transformation of the aerial warfare paradigm itself.

1.1.2.3 Characteristics of Manned-Unmanned Collaborative Operations

The significant differences in performance characteristics and operational modes between manned aircraft and UAVs introduce several distinctive features to their collaborative use in formations.

First, the formation concept has expanded substantially. Manned aircraft and UAVs differ greatly in performance and deployment. The traditional notion of homogeneous manned formations is no longer suitable for these novel hybrid formations. The external physical layout of the formation is no longer fixed; it can become more dispersed or even decentralized, operating beyond visual range, or the elements can be physically integrated, such as a UAV being carried by a manned aircraft during the early phase of a mission.

Second, hybrid formation configurations are highly flexible and diverse. The wide variety of UAV types (i.e., large/medium/small, long-/mid-/short-range, high-/mid-/low-altitude) enables them to form combat units with manned aircraft in numerous ways. These configurations can be flexibly tailored and used based on the specific requirements of the mission at hand.

Third, emphasis is placed on maximizing the comprehensive combat effectiveness of the entire formation. The hybrid formation does not prioritize the combat capability of any single platform but rather optimizes the integrated combat power of the entire formation unit.

Then, dense formations impose high demands on UAV autonomy and safety. Manned aircraft formations generally consist of the same or similar aircraft types with comparable performance, making it easier to maintain dense formations. Even in hybrid formations of different manned types, pilots can use tactical procedures to maintain relatively close formations. However, UAV operations remain dependent on control from manned platforms or ground stations. Conducting tight-formation, high-maneuverability combat operations or responding to sudden aerial threats places extremely high demands on their autonomous capabilities and operational safety assurance systems.

Lastly, UAV command authority and relationships must be strictly and explicitly defined. A UAV may fly based on a predefined mission profile, but it must rely on rear command and control for real-time situational changes. Typically, within a hybrid formation, a UAV might be subject to multiple control sources such as ground control stations and manned aircraft. Consequently, command authorities and relationships must be explicitly defined across all operational levels and scenarios for formation-based missions, as any ambiguity in these protocols would substantially degrade combat effectiveness.

1.1.2.4 Classification of Manned-Unmanned Cooperation Levels

In manned-unmanned hybrid formation operations, a primary consideration is the cooperation level, which defines the degree of control and autonomy exercised by the manned platform over the UAVs within the formation. This level is a critical factor influencing operational effectiveness and employment modes [9]. The U.S. Navy initially established definitions for these control levels, which were subsequently adopted by the STANAG 4586 standard and have been propagated across NATO member states for manned-unmanned cooperation programs [10–12]. This standard specifies five ascending control levels by which a manned platform can exercise authority over a UAV, with capabilities increasing from low to high.

(1) Level 1: There is no control relationship between the manned aircraft and the UAV. The manned aircraft indirectly acquires data from UAV sensors via the ground control station for display or distribution over

the data link network. The manned platform can receive data from ground stations through various existing communication systems, including V/U-band line-of-sight links, satellite communications, and tactical data links. This method requires minimal platform modification and is straightforward to implement.

(2) Level 2: The manned platform can directly acquire and display sensor data from the UAV via line-of-sight or beyond-line-of-sight links, eliminating the need for ground station relay or processing. This direct data transfer reduces communication latency and prevents potential information degradation associated with ground station processing, representing a significant improvement over Level 1 capabilities.

(3) Level 3: Building upon Level 2 capabilities, Level 3 enables the manned aircraft to control the UAV's payload and sensors, while the UAV operations remain managed by another platform. This tier thus necessitates two distinct control links to facilitate the real-time, separate management of the payload/sensors and the platform's flight.

(4) Level 4: In addition to the capabilities of Level 3, the manned aircraft possesses all control capabilities for the UAV, excluding takeoff and landing. Building upon payload control, it can command the UAV's flight path and perform obstacle and threat avoidance.

(5) Level 5: This level entails full functional control of the UAV from takeoff to landing. Regarding Level 1 collaboration, this equates to planned collaboration. Here, the manned aircraft and UAV operate independently under a unified mission plan, with no direct command-and-control relationship or information exchange between them. Regarding Level 2 collaboration, the primary function is sensor data reception. The manned aircraft can use a single link to receive target information from one UAV or employ multiple data links to simultaneously receive information from several UAVs. This suggests that at the same hierarchical level, application control can manifest in one-to-one, one-to-many, and many-to-one collaborative modes. Regarding Levels 3 to 5 collaboration, which involve control over the payload, flight, and takeoff/landing, the application control authority at a given hierarchical level can be held by only one platform at a time. When a platform possesses only a lower level of control authority, higher-level control is exercised by the UAV

ground control station via other data links. Within the U.S. military, existing UAVs such as the "Predator" and "Global Hawk" are capable of achieving Level 1 collaboration with manned aircraft. The VUIT-2 system installed on Apache helicopters enables simultaneous Level 2 control over multiple UAVs.

Overall, the current focus worldwide aims at achieving Level 4 control, employing a single manned aircraft to command one or multiple UAVs. Level 5 is generally reserved for large platforms such as transport and maritime patrol aircraft, where ground control station functionality may be retained.

1.2 DOMESTIC AND INTERNATIONAL DEVELOPMENTAL STATUS

1.2.1 Developmental Status of Manned-Unmanned Cooperative Operations

The U.S. Department of Defense's 2013 Unmanned Systems Integrated Roadmap considered manned-unmanned cooperation as a foundational operational mode following the military's strategic reorientation from the Middle East to the Asia-Pacific region. In fact, given the promising application prospects of hybrid manned-unmanned formations in collaborative operations, the United States, the United Kingdom, and other countries have been competing since the 20th century to conduct relevant research, exploring the feasibility, effectiveness, and practicality of manned-unmanned cooperation. In recent years, significant progress has been made, with breakthroughs in several key technologies, and the field has now entered the experimental validation phase [13–15].

As early as the flight tests of the X-45A UAV, the U.S. Air Force verified its airborne communication capability with a manned T-33 aircraft, representing the initial step in exploring hybrid formation and collaborative operations between manned and unmanned platforms. Subsequently, the U.S. Air Force initiated the Software Enabled Control (SEC) program, utilizing a T-33 (retrofitted as a UAV) and an F-15E to conduct flight tests for collaborative reconnaissance of high-risk area targets. These demonstrations successfully verified manned aircraft's control over UAV platforms [16, 17].

The U.S. Navy, in 2003, validated the control capability of the Tactical Control System (TCS) aboard a P-3C maritime anti-submarine patrol aircraft over the "Fire Scout" unmanned helicopter. In 2012, it then proposed the development of a Common Control System (CCS) program, aimed at achieving universal control over various types of UAVs, including unmanned reconnaissance aircraft and unmanned combat aerial vehicles.

Simultaneously, the U.S. Army also conducted a series of research and experiments on manned-unmanned cooperation. These efforts spanned multiple programs, including the Airborne Manned-unmanned System Technology, the Hunter Standoff Killer Team (HSKT), and the Manned-unmanned Common Architecture Program. In 2006, it successfully demonstrated the ability of a UH-60 "Black Hawk" helicopter to control a "Hunter" UAV and its imaging sensor through the Tactical Common Data Link (TCDL). Furthermore, flight tests were successfully conducted in which an "Apache" helicopter directly controlled a UAV's weapon launch. In this test, the "Apache" helicopter was linked via a data link to the sensor of an unmanned helicopter (modified from an AH-6 "Little Bird") and issued the command to launch a Hellfire missile, validating the collaborative combat capability between the helicopter and the UAV. Subsequent demonstrations expanded this capability, with the Apache helicopter successfully controlling diverse UAV platforms, including the "Hunter", "Shadow", "Raven", and "Gray Eagle", showing proficiency not only in intelligence acquisition but also in direct control of both the UAVs and their specialized payloads. In 2012, the first set of TCDL components for UAVs was delivered. This system enables Apache Block III crews to control the UAV's sensors and flight path, receive data transmitted from the UAV, reduce the sensor-to-shooter timeline, and significantly enhance the helicopter's capabilities in battlefield reconnaissance, surveillance, targeting, and mission execution [1].

For decades, the U.S. Army has been committed to developing high-performance weapon platforms, seeking military advantages far exceeding those of its competitors. However, despite their advanced capabilities, these high-performance platforms are extremely costly and suffer from prolonged development cycles, resulting in their electronic components being obsolete by the time they enter service. To address these challenges, the Defense Advanced Research Projects Agency (DARPA) approved and established the System of Systems Integration Technology and Experimentation (SOSITE) project [18]. This project focuses on distributed air combat concepts, architectures, and technology integration tools, aiming to maintain air superiority through a system-of-systems approach. It distributes aviation combat capabilities, including aircraft, weapons, sensors, and mission systems, across a large number of interoperable manned and unmanned platforms. This enhances the overall combat efficacy of multiple weapon systems and enables faster, lower-cost integration of the latest technologies and systems into existing aviation combat systems, as shown in Figure 1.1. Simultaneously, Lockheed Martin's Skunk Works proposed the "Fighter Engagement Manager (FEM)" concept, which represents a typical application of system-of-systems integration technology in air combat [19].

Figure 1.2 illustrates the system-integrated air-combat concept under the FEM framework, which utilizes one or more manned aircraft (F-22) teamed with multiple unmanned combat aerial vehicles (UCAVs) to collaboratively execute high-risk Suppression of Enemy Air Defenses/Destruction of Enemy Air Defenses (SEAD/DEAD) missions. The F-22 can command and control the UCAVs from outside the enemy's air defense zone and, if necessary, intercept aerial targets threatening the UCAVs. Experimental analysis indicates that the manned-unmanned cooperation model based on the FEM concept can increase the mission success rate by 35%, enhance combat efficiency by 25%, improve survivability by 25%, and reduce mission time by 50%.

The United Kingdom has conducted research on technologies enabling a single pilot to control up to five UAVs. In 2007, the Royal Air Force completed simulated manned-unmanned cooperative operation trials, in which a modified Tornado fighter aircraft commanded three UAV

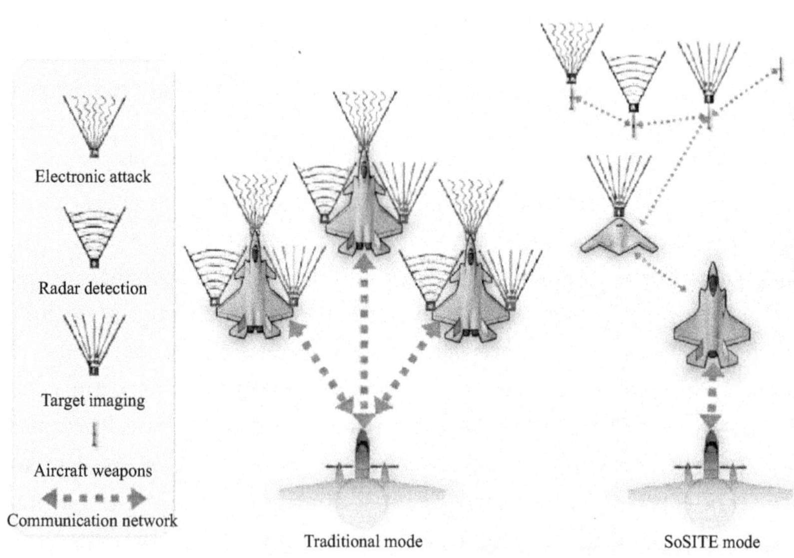

FIGURE 1.1 SoSITE project distributed air-combat concept diagram

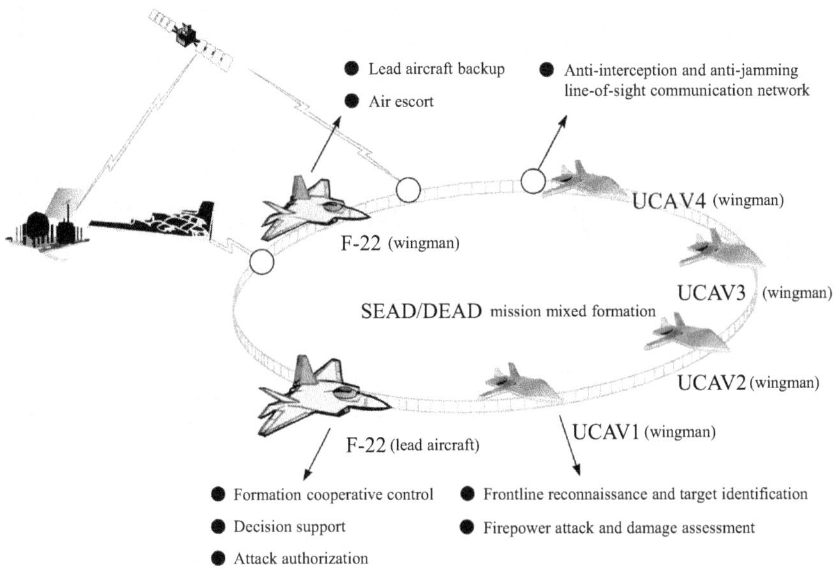

FIGURE 1.2 System-integrated air-combat concept diagram under the FEM concept

simulators (modified from BAC-111 aircraft) to conduct simulated attacks against moving ground targets. The UAVs autonomously performed all phases from takeoff to target detection, identification, and locking, while the final authorization for weapon release remained with the human pilot.

France has also been investigating manned-unmanned cooperative operations modes. According to reports, in March 2014, flight tests were conducted involving the Rafale fighter and the Neuron UAV, validating their capabilities for close formation flight and collision avoidance.

It is noteworthy that in June 2015, Dr. Endsley, the Chief Scientist of the U.S. Air Force, renowned for cognitive science research, stated in the report *Autonomy Horizon: The Path to Future Air Force Systems Autonomy*, "Conducting research on manned-unmanned cooperative operations based on human-autonomy teaming is an effective way to ensure the U.S. Air Force maintains a significant military advantage over its competitors" [20, 21]. One year later, the U.S. Air Force introduced the Loyal Wingman (LW) concept and conducted a demonstration flight of a manned-unmanned formation in March 2017, as shown in Figure 1.3 [22]. This concept aims to enable the pilot of a fifth-generation fighter (the lead aircraft) to control UAVs (wingmen) while simultaneously monitoring the

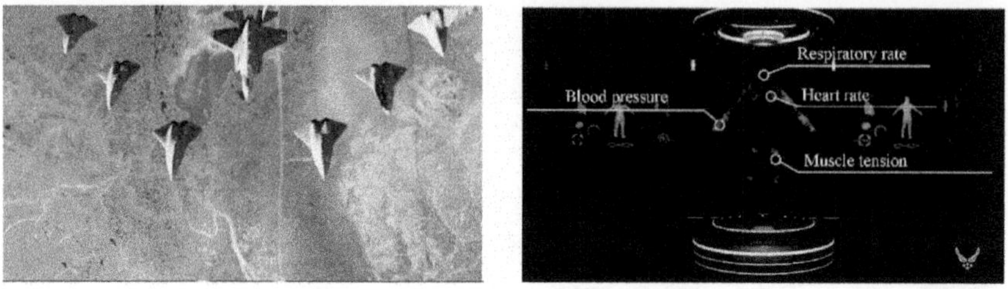

(a) Schematic diagram of the "Loyal Wingman" concept (b) Schematic diagram of pilot status monitoring

FIGURE 1.3 The "Loyal Wingman" concept: collaborative formation operations of an F-16 armed UAV with the F-35 fighter aircraft [23]

pilot's physiological status in real time, including respiratory rate, heart rate, blood pressure, and muscle tension, thereby enhancing the combat capabilities of both the lead aircraft and the wingmen. Such hybrid formations, integrating advanced manned aircraft with high-performance UAVs, leverage inherent complementary advantages in both platform characteristics (e.g., stealth, maneuverability, endurance, combat radius) and system capabilities (e.g., sensors, communications, computing, weapons). These formations are particularly well suited for executing demanding missions in contested environments such as Suppression/Destruction of Enemy Air Defenses (SEAD/DEAD). Consequently, they are regarded as a prioritized and operationally viable paradigm, representing a significant direction for the evolution of future combat methodologies [8].

At present, Western countries represented by the United States have conducted multiple demonstrations of manned aircraft and UAV collaborative concepts and comprehensive exercises of typical mission capabilities [24]. The autonomy level of current U.S. UAVs has reached Level III or above, and the autonomy level of UAVs under development has reached Level V. The main concepts and research projects are shown in Table 1.1.

Overall, extensive research has been conducted internationally in the areas of manned-unmanned cooperative operational modes, interoperability levels, collaborative "sensing-decision-making", cooperative control, and system integration. These investigations have achieved substantial depth and maturity within the field. Although current developments remain primarily driven by specific military branches and platform applications, a cross-branch, platform-agnostic, generalized manned-unmanned cooperation system has not yet been fully established, and the trend toward open, universal, and standardized systems is logically inevitable.

TABLE 1.1 Research on Manned-Unmanned Cooperative Operations in the United States

Serial Number	Concept/Research Project Name	Supervisory/ Implementation Unit	Research Content
1	Fighter engagement management	U.S. Air Force Research Laboratory and Lockheed Martin Corporation	Study on the formation composed of two F/A-22 aircrafts and four small multi-role UCAVs, achieving cooperative execution of high-risk and challenging reactive SEAD/DEAD missions.
2	Loyal Wingman	U.S. Air Force Research Laboratory and Lockheed Martin Corporation	Study on the open-system architecture for formations of unmanned-converted F-16 and F-35, and their capabilities for executing autonomous mission planning, SEAD/DEAD operations, and contingency handling.
3	Software-enabled control	Defense Advanced Research Projects Agency	Study on the capability of an F-15E to control a T-33-converted unmanned aircraft via voice, enabling interoperable route planning, search, and strike operations.
4	Airborne manned/ unmanned systems technology	The Boeing Company	Study on the cooperative capability of one AH-65D armed helicopter and one "Hunter" UAV to operate collaboratively via an intelligent video-sharing system.
5	Manned/unmanned aircraft common architecture plan	U.S. Army	Study on the cooperative interoperability capabilities between multiple helicopter types (e.g., Apache, Kiowa) and multiple unmanned aircraft (e.g., Shadow, Fire Scout).
6	UAV collaborative operations in denied environments	Defense Advanced Research Projects Agency	Study on the adoption of the RQ-23 "TigerShark" to enable manned aircraft to command UAV formations for conducting tactical reconnaissance, SEAD/DEAD, and related missions.

1.2.2 Developmental Status of Manned-Unmanned Supervisory Control Technology

1.2.2.1 Research on Single Operator Control of Multiple UAVs

The number of UAVs a single operator can simultaneously control is largely determined by the level of autonomy (LOA) of the UAVs in three control loops, including flight control, navigation, and mission and payload management, as shown in Figure 1.4 and Table 1.2.

In domestic and international research on multi-UAV supervisory control capabilities, researchers have consistently delegated the lowest-level flight control loop to the UAV autonomous systems, thereby freeing up more human cognitive resources for higher-level control loops. The research findings primarily indicate the following:

(1) Enhancing autonomy across the three control loops can improve a single operator's capacity to manage multiple UAVs concurrently. However, this capability is also constrained by factors such as system reliability and trustworthiness, communication bandwidth, mission complexity, operational tempo, and the adequacy of operator training.

(2) While high autonomy levels can improve operational efficiency, they may also foster operator overreliance on autonomous systems, potentially leading to complacency and decision-making errors. Maintaining adequate operator situational awareness remains crucial to promptly identify system anomalies and implement necessary interventions.

(3) The scalability of human supervision in UAV operations is directly influenced by the frequency and intensity of human-machine interactions. For instance, a single operator has demonstrated the capacity to simultaneously manage up to 12 cruise missiles, a scenario characterized by minimal interaction demands. In this configuration, the operator primarily monitored the mission management loop, while all other control functions operated autonomously [26].

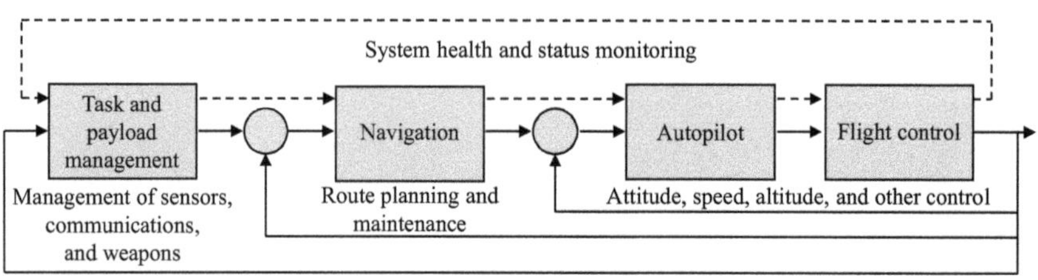

FIGURE 1.4 Hierarchical control loops of a single UAV

TABLE 1.2 Autonomy Level Classification and Description

LOA	Improved LOA	Autonomy Level Description
1	I	The computer-intelligent component system provides no assistance, and the operator must make all decisions and carry out all actions manually.
2	II	The computer-intelligent component system provides all possible solutions for decisions and actions.
3	III	The computer-intelligent component system provides an optional subset of decisions and actions.
4/5	IV	The computer-intelligent component system provides a decision or action plan, which is executed upon operator approval, and this is referred to as the consent management mode.
6	V	The computer-intelligent component system provides a decision or action plan, which the operator may veto within a specified time window before execution. Otherwise, the system automatically carries out the decision or action. This mode is referred to as the exception management mode.
7/8/9/10	VI	The operator does not participate in the decision-making process, and the computer-intelligent component system makes and executes decisions autonomously.

Source: [25] Xu, J. (2017). *Autonomous Ground Attack Decision-Making of UCAV Based on FCM Model Theory* [Master thesis, Northwestern Polytechnical University].

In summary, a broad consensus exists within the international defense community that, when UAV flight control is fully automated, navigation systems employ artificial intelligence to achieve Level IV-V autonomy, and mission/payload management operates under consent-based protocols (Level IV), a single operator in a manned aircraft can effectively control four to five UAVs simultaneously in standard SEAD/DEAD missions.

1.2.2.2 Research on Multiple UAV Supervisory Control Technology

In recent years, multi-UAV supervisory control technology has gained significant attention worldwide. The design of supervisory control systems focuses on optimizing human-machine interaction by selecting appropriate autonomy levels based on operational dynamics. This approach establishes effective interactive paradigms that balance human oversight with system autonomy. The U.S. Air Force Research Laboratory (AFRL) evaluated the performance of human-machine systems across multiple autonomy levels. Analyses demonstrated that autonomous unmanned systems could assist operators in completing missions, with different autonomy levels impacting the overall effectiveness of human-machine systems. AFRL also proposed a variable autonomy framework for UAV supervisory control, where variable autonomy supervisory control technology and mixed-initiative human-machine interaction technology serve as key implementation methods. In an AFRL-designed simulation experiment, a single operator controlled four unmanned combat aerial vehicles performing SEAD/DEAD missions while simultaneously measuring the operator's physiological records. These physiological metrics were employed to evaluate the operator's cognitive state, facilitating real-time allocation of control authority between the human operator and the UAV's autonomous systems according to both cognitive workload and mission demands.

In supervisory control human-machine interaction, situation awareness (SA) presents an unavoidable yet challenging issue [27]. Endsley provided the widely accepted definition of situation awareness, that is, "the perception of environmental elements within a volume of time and space, the comprehension of their meaning, and the projection of their status in the near future" [28]. In 2015, Endsley synthesized nearly 25 years of research in SA, addressing questions and critiques of the model originally proposed in 1995. The study systematically compared newer SA frameworks, including situated, distributed, and sense-making approaches, with established individual and collective models, examining both their distinctions and interrelationships. Overall, the research community has gradually reached the following consensus:

(1) Human-machine interaction characteristics and limited-intervention features are key scientific issues in supervisory control research.

(2) Variable autonomy is not the goal of human-machine interaction but an important means of supporting it.

(3) Situation awareness significantly influences supervisory control effectiveness and serves as a critical basis for variable autonomy adjustments.

(4) Workload and situation awareness levels affect the operator's cognitive state, which can be measured using physiological data such as eye movement and heart rate, or evaluated or predicted through task load models.

1.2.3 Developmental Status of Manned-UAVs Collaborative Decision-Making Technology

Analysis of the U.S. military's operational concepts and research projects listed in Table 1.1 reveals distinct trends in cognition and autonomy. Cognition in contested environments is generally demonstrated through comprehension, judgment, and other cognitive activities, along with the cognitive states of pilots in manned aircraft. These elements represent critical factors in the collaborative decision-making process. Advanced autonomy is demonstrated through newer-generation UAVs achieving higher levels of independence, equipped with inherent capabilities for situation assessment and intelligent reasoning in contested environments. This technological progression establishes the foundational framework for evolving human-machine collaboration from master-slave relationships to peer-to-peer cooperative decision-making.

Human-machine interaction serves as the foundation for effective collaboration between manned aircraft and UAVs. Adopting multimodal human-machine interface design methods can effectively reduce pilots' cognitive workload. Incorporating human factors and ergonomics (HFE) into the design of human-machine interaction systems represents the future development direction for manned-unmanned vehicle interaction technologies. This requires analyzing issues such as different interactive modes, interaction complexity, and criticality, while proposing a novel human-machine interaction framework from the perspective of cognitive intelligent interaction.

At present, supervisory control emphasizing "human-in-the-loop" and "human-on-the-loop" represents a critical mode for achieving human-machine collaborative decision-making. However, in current research, UAVs predominantly assume a subordinate role, with insufficient decision-making initiative. The capability for SA is crucial for UAVs to bridge the cognitive gap with humans, advance from intelligent to autonomous decision-making, and achieve peer-level collaborative decision-making with human operators. Situation awareness aligns with the observe and orient phases of the OODA loop, serving as the foundational element for both human cognitive decision-making and autonomous UAV decision processes. This cognitive theory supports decision-making in complex, dynamic environments through two complementary approaches, top-down task-driven and bottom-up data-driven processes, both guided by mental models. Implementing human-like mental models in UAVs is essential for achieving operational SA.

Overall, research in this field has gradually formed the following consensus:

(1) Based on UAV autonomous intelligence levels and operator intervention requirements, collaborative decision-making between manned aircraft and UAVs can be categorized into limited intervention at basic collaboration levels and cognitive-intelligent interaction at advanced collaboration levels.

(2) Effective interaction between human "cognition" and UAV "autonomy" is the foundation for advanced collaboration between future sophisticated manned aircraft and high-end UAVs, and it is key to achieving collaborative decision-making between high-level and low-level intelligent entities.

(3) In dynamic and uncertain, complex environments, constructing a complete OODA loop that incorporates autonomous interaction strategies, autonomous emotion, SA, and intelligent decision-making is an effective approach to achieving autonomous decision-making in UAVs.

(4) In the process of collaborative decision-making between manned aircraft and UAVs, the influence of pilots' physiological and psychological factors on cognitive decision-making must be considered, as this is a prerequisite for determining whether research on collaborative decision-making can be translated into practical applications.

(5) Expert knowledge and experience are important foundations for building explainable military decision-making systems. However, the ability to learn and evolve with available data will enhance the adaptability of collaborative decision-making systems in practical applications.

1.2.4 Developmental Status of Fuzzy Cognitive Map Model Theory

Conducting research on collaborative decision-making between manned aircraft and UAVs requires seeking a set of feasible analytical methods and modeling tools based on theories such as information science, control science, and cognitive science. Among these, fuzzy cognitive maps (FCM) represent a relatively ideal tool for modeling and simulating dynamic system knowledge representation and decision support [29]. According to statistics from the Web of Science Core Collection (as of March 2020), the number of related research papers reached 115, 120, and 100 in 2017–2019, respectively. In 2019 alone, these papers were cited 2,351 times, with a total citation frequency of 15,637 times and an h-index of 58, establishing FCM as an important research direction in the field of artificial intelligence.

FCM is a qualitative reasoning technique and soft computing method. The classical FCM model consists of concept nodes, directed arcs, and their associated weight matrix. Nodes represent system attributes, characteristics, and performance. Directed arcs represent influence relationships between nodes, with the degree of influence described by the associated weight matrix. The entire FCM simulates complex system behavior through causal relationships and interactions among concept nodes, as shown in Figure 1.5. The FCM architecture enables the mapping of input states to convergent outputs, which may manifest as either fixed points or limit cycles, though the system can also enter a divergent state characterized by chaotic dynamics.

Because the FCM is a simple and flexible modeling approach, with ease of integrating other intelligent methods and learning algorithms, along with strong adaptability and usability across diverse application scenarios, FCM models have developed rapidly and given rise to numerous extended FCM architectures [30]. These primarily include Rule-Based Fuzzy Cognitive Maps [31], Fuzzy Gray Cognitive Maps [32], Intuitionistic Fuzzy Cognitive Maps [33], Dynamic Fuzzy Cognitive Maps/Cognitive Networks [34], Fuzzy Timed Cognitive Maps [35], Evolutionary Fuzzy Cognitive Maps [36], and Agent-Based Fuzzy Cognitive Maps [31]. Research on FCM learning focuses on learning the weight matrix of FCMs, including node relationships and their strengths [37]. There are other methods, including Hebbian-based learning [38], population-based learning [39], and hybrid

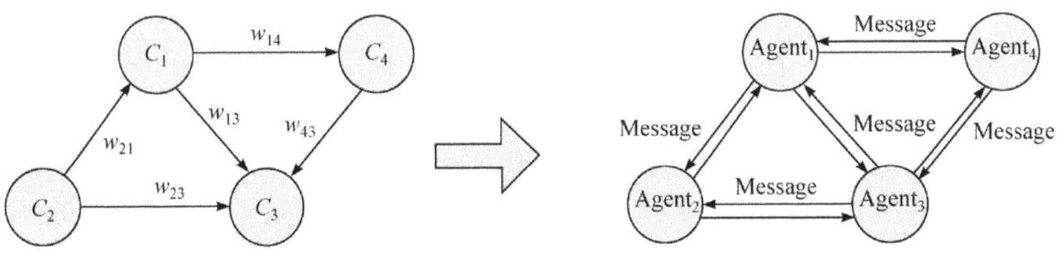

FIGURE 1.5 The mapping relationship between classical FCM and agent-based FCM

learning [40]. These have been applied to modeling, analysis, and decision support in numerous fields such as society [33], environment [41], industry [34], and healthcare [30, 38, 42].

The following section briefly introduces several extended models and learning algorithms with promising potential in decision system applications.

1.2.4.1 Agent-Based Fuzzy Cognitive Map

In the Agent-Based Fuzzy Cognitive Map (ABFCM), each node can be an agent of varying forms and structures, equipped with distinct reasoning algorithms. These nodes interact with one another through a message-passing mechanism. When the algorithm for each node is consistent, the system reverts to the classical FCM. ABFCM is suitable for modeling distributed decision-making systems, as illustrated in Figure 1.5.

Equation (1–1) represents the reasoning model for node state values in classical FCM:

$$C_j(t+1) = f\left[k_1 \sum_{\substack{i=1 \\ i \neq j}}^{n} C_i(t) w_{ij} + k_2 C_j(t)\right] \tag{1–1}$$

where $C_i(t)$ $C_j(t)$ represent the state values of the cause node and effect node at time t; w_{ij} denotes the causal weight from the concept node C_i to C_j; $k_1, k_2 \in [0,1]$ indicate the proportional contributions of the integrated influence from cause nodes and the node's own state; f represents the activation function of concept nodes, which can be expressed in various forms [43].

ABFCM maps the node C in classical FCM to an agent A, transforming the node state reasoning equation from Equation (1–1) into the form of Equation (1–2):

$$A_i^t = F_i(A_1^{t-1}, \cdots, A_j^{t-1}, \cdots, A_n^{t-1}, w_{11}, \cdots, w_{ji}, \cdots, w_{nn}) \tag{1–2}$$

where A_i^t represents the state of the i agent at time t; w_{ji} denotes elements in the weight matrix, and F_i refers to the reasoning process through which the state is A_i^t obtained based on the given information, replacing the activation function used in classical FCM.

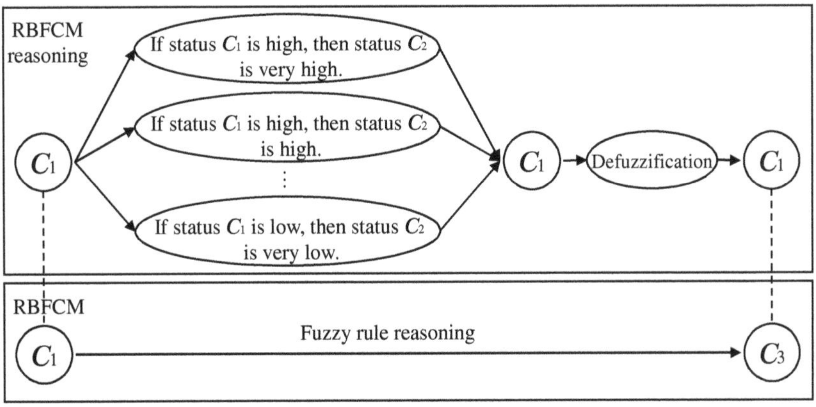

FIGURE 1.6 RBFCM and its inference relationships

1.2.4.2 Rule-Based Fuzzy Cognitive Map

In the Rule-Based Fuzzy Cognitive Map (RBFCM), the relationships between concept nodes, originally expressed through fuzzy values, are described using fuzzy rules. This approach can represent not only the monotonic causal relationships of classical FCMs but also various interrelations such as reasoning, substitution, probability, opposition, and correlation. The activation function is replaced by fuzzy inference rules, enabling the modeling of more complex real-world decision-making systems, as illustrated in Figure 1.6.

1.2.4.3 Dynamic Fuzzy Cognitive Map

As shown in Figure 1.7, the Dynamic Fuzzy Cognitive Map (DFCM) primarily introduces a set of external environmental control variables V, transforming originally static numerical values in the model into dynamically adjustable functions dependent on the initial values and V. These variables can directly modulate causal relationships between concept nodes, while simultaneously influencing node states and their fuzzification processes. This capability enables the reasoning mechanism to effectively incorporate external environmental factors into the decision-making system.

1.2.4.4 Fuzzy Gray Cognitive Map

The Fuzzy Gray Cognitive Map (FGCM) incorporates gray system theory into the FCM framework, as illustrated in Figure 1.8. It is defined as a

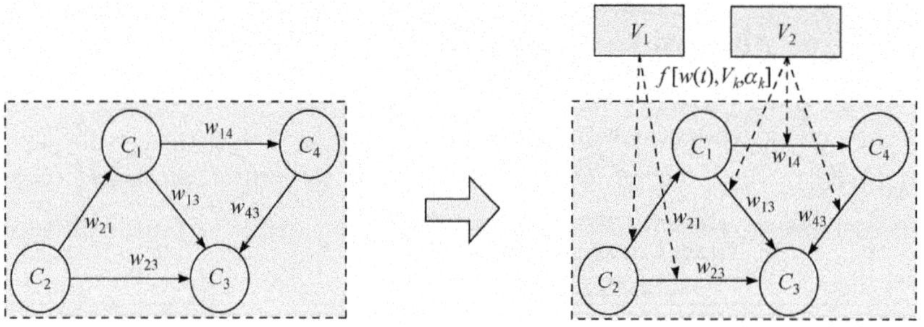

FIGURE 1.7 Extension relationship between classical FCM and DFCM

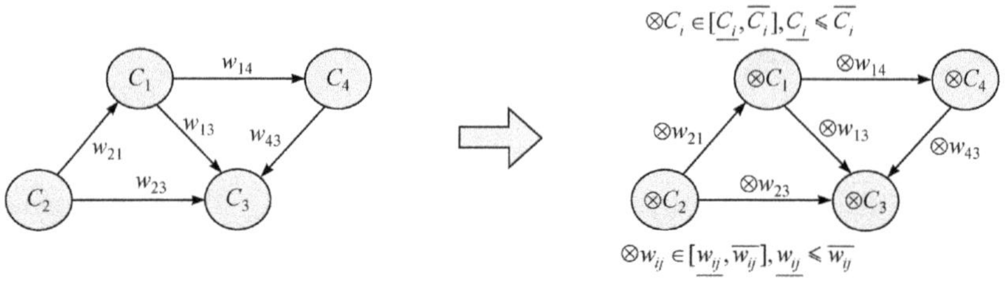

FIGURE 1.8 Extension relationship between classical FCM and FGCM

quadruple $\{C, W, \otimes f(\cdot), l(\psi)\}$, where $C = \{\otimes C_i\}_{i=1}^{n}$ represents the set of concept nodes with gray values. $W = \{\otimes w_{ij}\}$ denotes the magnitude of connections between concept nodes, also characterized by gray values. $\otimes f(\cdot)$ refers to the activation function of concept nodes. $l(\psi)$ defines the information space scope of the nodes. Converting gray values to crisp values requires a whitening process. Compared to classical FCM, FGCM uses gray values to represent both concept nodes and weights, enabling comprehensive expression of uncertainties in decision-making systems.

1.2.4.5 Learning Methods for Fuzzy Cognitive Maps

(1) Hebbian-Based Learning. The core concept involves adjusting and updating weights based on the sequence, manner, and intensity of concept node activation. Weights decrease for asynchronously activated nodes and increase for synchronously activated nodes. Key algorithms include Differential Hebbian Learning (DHL), Active Hebbian Learning (AHL), Nonlinear Hebbian Learning (NHL), and Data-Driven Nonlinear Hebbian Learning (DD-NHL).

(2) Population-Based Learning. The core idea involves employing population-based heuristic algorithms to train FCMs. Main methods include Evolutionary/Genetic Algorithms (EA/GA), Ant Colony Optimization (ACO) algorithm, Particle Swarm Optimization (PSO) algorithm, Chaotic Simulated Annealing (CSA) algorithm, Tabu Search, and Evolutionary Strategies.

(3) Hybrid Learning. The core idea is to learn, modify, and update the weight matrix of FCMs in two stages, based on initial domain expert knowledge and available historical data. Main approaches include hybrid methods combining Hebbian Learning with Differential Evolution, hybrid methods combining Hebbian Learning with the Extended Great Deluge algorithm, and hybrid methods using Real-Coded Genetic Algorithms.

Table 1.3 summarizes the advantages, disadvantages, and applicable scopes of commonly used extended FCM models and learning methods. These FCM extensions and learning approaches have laid a solid foundation for

TABLE 1.3 Summarization of the Advantages, Disadvantages, and Potential Scopes of Commonly Used FCM Extensions and Learning Methods

Model/Method	Advantages	Disadvantages	Scope of Applications
ABFCM	Support for multi-agent node modeling	High requirements for modeling	Distributed decision-making
RBFCM	Support fuzzy rule modeling and inference	Reasoning process is complex	Expert decision-making
DFCM	Support nonlinear dynamic function inference	High requirements for modeling	Dynamic decision-making
FGCM	Support uncertainty representation of the model	Reasoning process is complex	Uncertain decision-making
Hebbian-based learning	Support unsupervised learning and converges quickly	Relying on expert knowledge, and low accuracy	Lack of available learning data
Group-based learning	Support supervised learning with high accuracy	Slow convergence and poor interpretability	It has a large amount of learning data
Blended learning	Support online and offline step-by-step learning	The algorithm is complex and difficult	Knowledge and data dual-drive

addressing key challenges in complex military decision-making systems, including framework design, system modeling, and the representation and reasoning of dynamic characteristics and uncertainties, as well as leveraging historical data and expert knowledge for learning. However, applying FCM to research on manned-unmanned collaborative decision-making modeling methodologies still requires adaptive modifications and integrated application.

CHAPTER 2

Analysis of the Limited-Intervention Collaborative Decision-Making Mechanism between Manned Aircraft and UAVs

THIS CHAPTER INTRODUCES THE concepts of limited intervention-based autonomous decision-making and autonomy levels for UAVs. It provides an analysis of four key mechanisms, including the hierarchical intelligent autonomous control and decision-making mechanism for UAVs, the limited-intervention mechanism for manned aircraft, the autonomy level adjustment mechanism for UAVs, and the operational scheduling mechanism for human machine collaborative decision-making.

2.1 PROBLEM FORMULATION OF LIMITED INTERVENTION-BASED AUTONOMOUS DECISION-MAKING FOR UAV

Autonomy refers to the capacity for self-governance and self-directed decision-making. For UAVs, autonomy is defined as the ability to sustain

DOI: 10.1201/9781003791713-2

mission completion in uncertain and complex environments without human intervention.

The prerequisite for achieving autonomous decision-making in UAVs is a high level of autonomy, and the required modes of interaction between the operator and the unmanned system vary significantly across different autonomy levels. Limited intervention-based collaborative decision-making operates under a supervisory control paradigm, wherein the manned aircraft operator is extensively involved in higher-level payload and mission management of the unmanned system. The "permission management mode" and the "exception management mode" are two common supervisory control methods. Figure 2.1 illustrates the structure of limited intervention-based autonomous decision-making for UAV [44].

The autonomous decision-making process of UAVs can be divided into three layers: (1) The deliberative reasoning layer has the highest degree of autonomy, and it derives SA outcomes through reasoning based on acquired situational information. (2) The decision-making layer utilizes the situational cognition from the upper layer to conduct mission planning and selects corresponding action sequences to output mission planning results. (3) The operational execution layer has the lowest degree of autonomy and is responsible for action execution. The limited-intervention

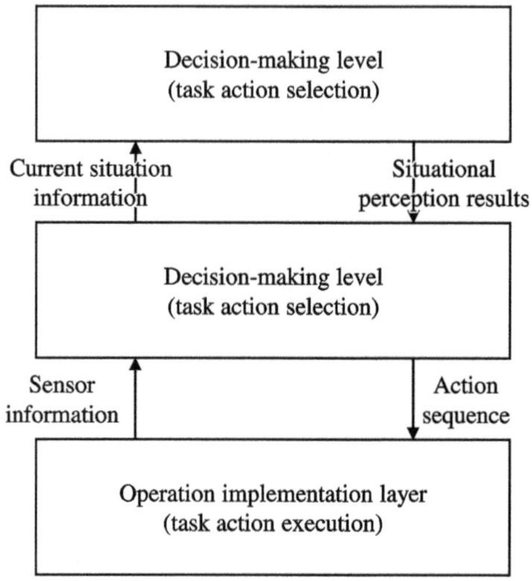

FIGURE 2.1 Architecture for limited intervention-based autonomous decision-making in UAVs

mechanism dynamically allocates execution authority for each layer to the manned aircraft operator and the UAV according to different autonomy levels during the mission execution process. For example, under the fully autonomous condition, the UAV possesses control authority over the deliberative reasoning layer. Under the remote operational mode, both SA and reasoning in the deliberative reasoning layer and task action selection in the decision-making layer are completed by the operator, while the UAV is responsible for action execution. The selection of the autonomy level for the UAV is related to the battlefield environment in which the manned aircraft and UAV formation operate. The role of human operators in the autonomous decision-making process of UAVs is shown in Figure 2.2.

In this process, the UAV can proactively report its current status and request assistance from the operator to complete certain tasks. Simultaneously, the operator can acquire situational information through the unmanned system and actively intervene in its decision-making process based on cognitive outcomes, thereby altering its autonomy level and mission execution methods. The realization of this operational framework necessitates research across three critical dimensions:

(1) Autonomy level classification for UAVs based on limited intervention

(2) Establishment of a hierarchical autonomous decision-making model framework for UAVs

(3) Design of limited-intervention strategies and intervention methods for manned aircraft

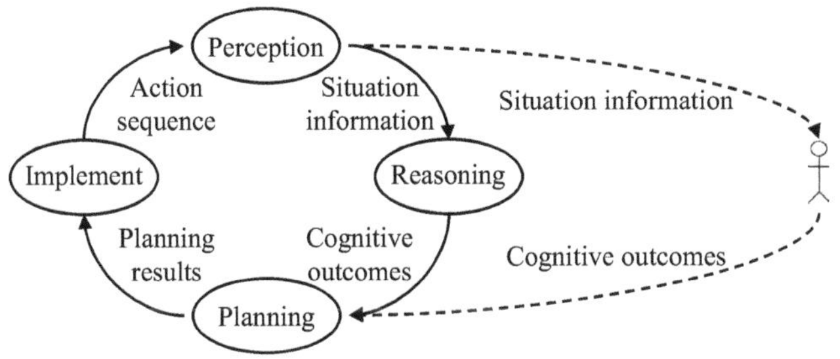

FIGURE 2.2 The role of human operators in the autonomous decision-making process of UAVs

2.2 AUTONOMY LEVELS FOR UAV BASED ON LIMITED INTERVENTION

The limited-intervention approach to classifying UAV autonomy uses broader categories than those shown in Table 2.1. This simpler system reduces the need for frequent adjustments to autonomy levels and is easier to apply in real-world operations. This classification primarily comprises exception management (corresponding to Level V in the modified LOA), permission management (Level IV), operator decision-making (Levels II and III), and directive control (Level I). Research indicates that operators generally prefer the permission management mode under normal circumstances. However, when facing time pressure, high workload, or particularly complex yet low-risk missions, operators tend to favor the exception management mode [45–52]. Manual remote control of the UAV remains the most fundamental and reliable operation method. Given that UAVs will continue to require operator guidance for the foreseeable future, Level VI (full autonomy) of the modified LOA is omitted. Consequently, the exception management mode is designated as the highest autonomy level for UAVs within this framework. The definitions of the autonomy levels for UAVs based on limited intervention are detailed in Table 2.1 [53].

The exception management mode represents an operational state where the UAV demonstrates high confidence in its situational assessment. In

TABLE 2.1 Definition of Autonomy Levels for UAVs Based on Limited Intervention

Autonomy Level	Interactive Content	Interaction Methods
A_4 (exception management)	UAV provides task execution status and system status.	UAV provides proactive feedback to the operator; if no objection is received within the specified time limit, the operation proceeds automatically.
A_3 (agree to management)	UAVs provide SA and mission decision results, which operators can approve or modify.	UAV waits for the operator feedback before reducing the action level.
A_2 (operator decision)	UAVs provide SA results and mission parameters, enabling operator decision-making.	UAV awaits the operator's decision before attempting to protect itself.
A_1 (command control)	UAV provides SA parameters, and the operator provides task instructions.	UAV slows down and awaits operator instructions.

this mode, the system autonomously completes SA and mission planning, then proceeds to execute the planned actions unless explicitly countermanded by the operator within a designated timeframe. The permission management mode refers to a state where the UAV perceives a degree of uncertainty in the current situation, and it requires explicit approval from the operator before it can execute any planned actions. The operator decision-making mode refers to when the UAV is unable to adequately assess the current situation, and it requires the cognitive assistance of the operator to complete the mission planning process. The directive control mode is defined as an operational state in which the UAV is incapable of effectively responding to the current situation. In this mode, all actions are executed solely based on direct commands from the operator.

2.3 UAV'S HIERARCHICAL AUTONOMOUS DECISION-MAKING FRAMEWORK

The hierarchical autonomous decision-making framework for UAVs primarily comprises three layers, including the SA layer, the decision-making and reasoning layer, and the human-machine interaction layer. These layers are respectively responsible for a UAV's perception of external environmental information, autonomous decision-making and reasoning, and human-machine interaction between the manned aircraft and UAVs. Figure 2.3 illustrates the hierarchical autonomous decision-making model for UAVs.

2.3.1 Situational Awareness Layer

The SA layer primarily consists of three modules, including an external information integration module, a node status module, and a situational reasoning module. The external information integration module periodically acquires external data, synthesizing environmental information, control commands or intervention instructions from the manned aircraft, and outcomes from situational reasoning to form external comprehensive information. The node status processing module extracts key factors representing the current state of the UAV as conceptual nodes, which are used for situational reasoning and subsequent decision-making inference. The situational reasoning module utilizes the integrated external information and node status data to derive situational cognition results.

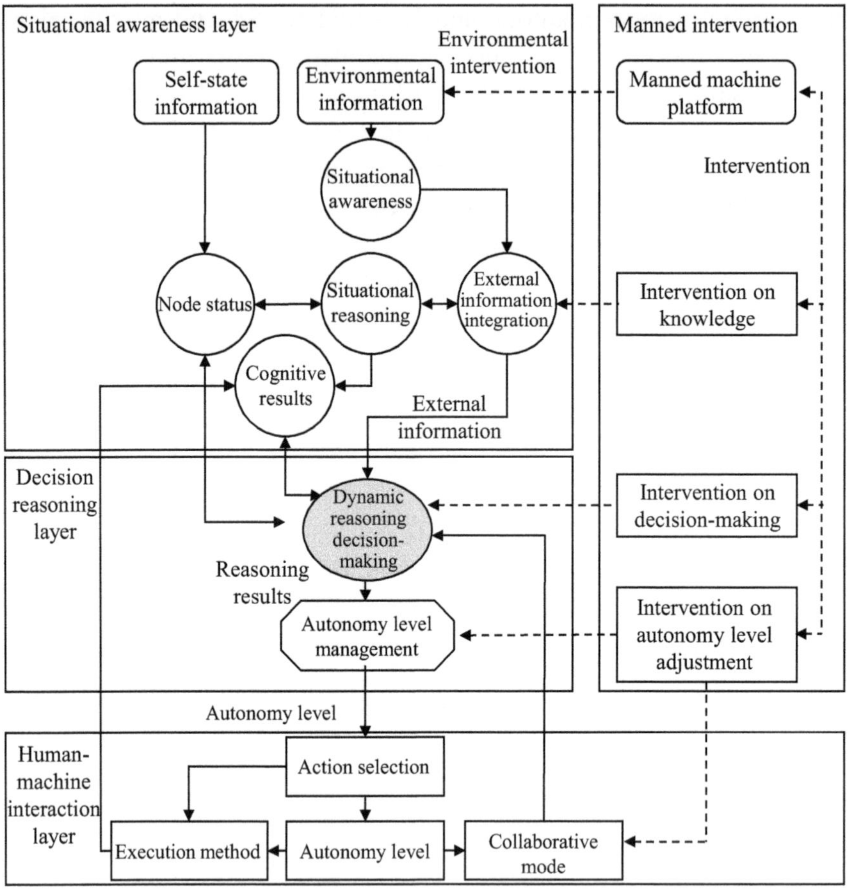

FIGURE 2.3 Hierarchical autonomous decision-making framework for UAVs

2.3.2 Decision-Making and Reasoning Layer

The decision-making and reasoning layer primarily receives preprocessed and integrated external environmental information sets and node statuses from the upper layer for dynamic inference and decision-making. It mainly consists of the dynamic inference and decision-making module and the autonomy level management module. The dynamic inference and decision-making module, serving as the core of this layer, is primarily responsible for generating inference results and performing decision reasoning. It simultaneously updates the node status and its internal knowledge base in real time. The inference results include task planning parameters and task planning outcomes. The autonomy level management module evaluates and adjusts the autonomy level based on the autonomous inference results, combined with the UAV's knowledge and operational permissions, thereby constraining the range of available actions. The autonomy level can be adjusted either proactively by the UAV itself or through intervention by the manned aircraft.

2.3.3 Human-Machine Interaction Layer

The human-machine interaction layer serves as the interface through which the UAV acts upon the external environment and, simultaneously, as the point of interaction between the manned aircraft and the UAV. Based on the inference results from the decision-making and reasoning layer and the autonomy level of the UAV, it selects and executes actions corresponding to that level from an action library. The autonomy level determines the mission execution method of the UAV, thus defining the collaborative mode between the manned aircraft and the UAV. Furthermore, this layer, with the external environment, influences the subsequent cognitive and reasoning processes of the UAV.

2.3.4 Intervention from the Manned Aircraft Platform

The dashed lines in Figure 2.3 represent limited intervention from the manned aircraft toward the UAV, which manifests to varying degrees across all three layers of the UAV's autonomous decision-making hierarchy. The intensity of limited intervention escalates from weak to strong as follows: intervention on the environment, intervention on knowledge, intervention on decisions, and intervention on autonomy level adjustment. These can be specifically categorized into three main types: (1) Intervention on the UAV's environment: This primarily refers to the influence exerted by the manned aircraft on the UAV mission environment and the modification of situational conditions. (2) Intervention on the UAV's knowledge: Through cognitive judgment and information provision, the manned aircraft alter the knowledge volume of the UAV. (3) Intervention on the UAV's decision-making and autonomy level adjustment: This changes the collaborative mode between the manned aircraft and the UAV and alters the reasoning and decision-making processes of the UAV. Intervention triggering occurs in two forms: intervention initiated by the UAV's active request and intervention proactively initiated by the manned aircraft.

2.4 LIMITED-INTERVENTION TRIGGERING STRATEGIES AND ENGAGEMENT METHODS FOR MANNED AIRCRAFT

The limited-intervention mechanism for manned aircraft primarily comprises two key components: intervention triggering strategies and intervention engagement methods. The implementation of limited intervention necessarily leads to adjustments in the autonomy level of the UAV, thereby resulting in the transfer and reallocation of decision-making authority between the manned aircraft and the UAV. In designing the limited-intervention mechanism, the operator's mental workload in the manned aircraft must be fully considered [54,55].

2.4.1 Limited-Intervention Triggering Strategies

Limited-intervention triggering strategies primarily depend on the autonomy level of the UAV, the task load and mental workload of the manned aircraft operator, and the complexity of the battlefield environment. Generally, when the UAV operates at a low autonomy level and the manned aircraft operator experiences a light task control load, the manned aircraft proactively initiates intervention. Conversely, when the UAV operates at a high autonomy level and the manned aircraft operator faces a heavy task control load, the intervention request is initiated by the UAV itself. The need for manned aircraft intervention in the autonomous decision-making process of the UAV arises from several factors, such as insufficient knowledge or resources to handle the current situation, a lack of functional capability to execute specific actions at the current autonomy level, or the absence of authority to perform certain actions within the current autonomy level framework. The triggering of intervention is always accompanied by an adjustment to the autonomy level of the UAV.

Figure 2.4 illustrates the limited-intervention triggering strategy. The dashed rectangular box on the left-hand side represents the autonomous decision-making process of the UAV, while the dashed rectangular box on the right-hand side represents the limited-intervention process of the manned aircraft. From the perspective of the UAV, intervention can be requested from the manned aircraft at any stage of its autonomous decision-making process. Conversely, from the perspective of the manned aircraft, intervention can be triggered either upon receiving a request from the UAV or through proactive initiation. In the autonomous decision-making process of the UAV, the autonomy level increases through a bottom-up mechanism, with decision behaviors becoming more complex and the decision process becoming more autonomous. In the intervention framework of

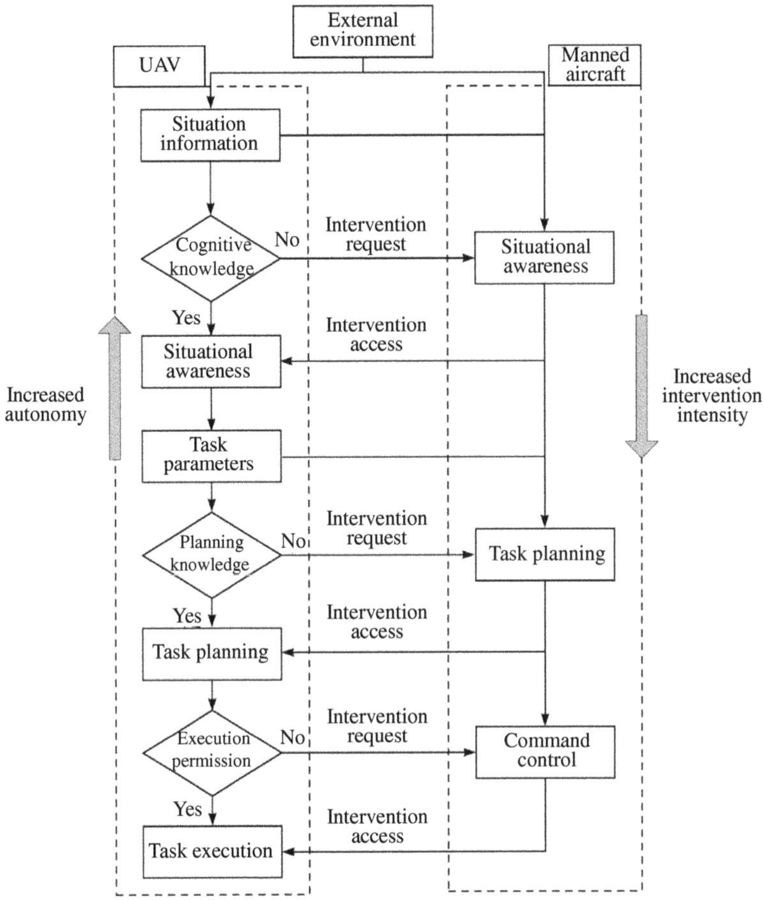

FIGURE 2.4 Limited-intervention triggering strategy

the manned aircraft, the intensity of intervention increases through a top-down mechanism, requiring increased allocation of attention and cognitive resources from the operator, consequently raising the workload.

Intervention by the manned aircraft at any specific layer implies that decision-making authority for that layer is transferred from unmanned to the manned aircraft, resulting in a corresponding decrease in the autonomy level of the unmanned system. For example, intervention at the situational cognition level constitutes fine-tuning of the UAV's autonomy level. Intervention at the mission planning level will alter the decision-making and reasoning behavior of the UAV. Intervention at the task execution level will prompt the operator to assume majority control authority, with the UAV executing tasks incrementally according to operator commands.

2.4.2 Limited-Intervention Engagement Methods

Figure 2.5 illustrates the engagement methods for limited intervention by manned aircraft. These methods can be categorized into four forms: direct control, environmental intervention, knowledge intervention, and decision intervention. When the manned aircraft operator is under a heavy task load, assistance is provided to the UAV in adjusting its autonomous decision-making and reasoning behavior. This is achieved through the manned aircraft's intervention in the operational environment, which modifies the UAV's SA and the initial state of its reasoning nodes. If the UAV remains unable to complete decision-making under these conditions, knowledge supplementation can be employed to modify the knowledge base supporting its autonomous decision-making processes, thereby enhancing its global SA. This approach leverages the cognitive analytical capabilities of the manned aircraft operator to assist the UAV in completing its decision-making. When the intervention methods mentioned

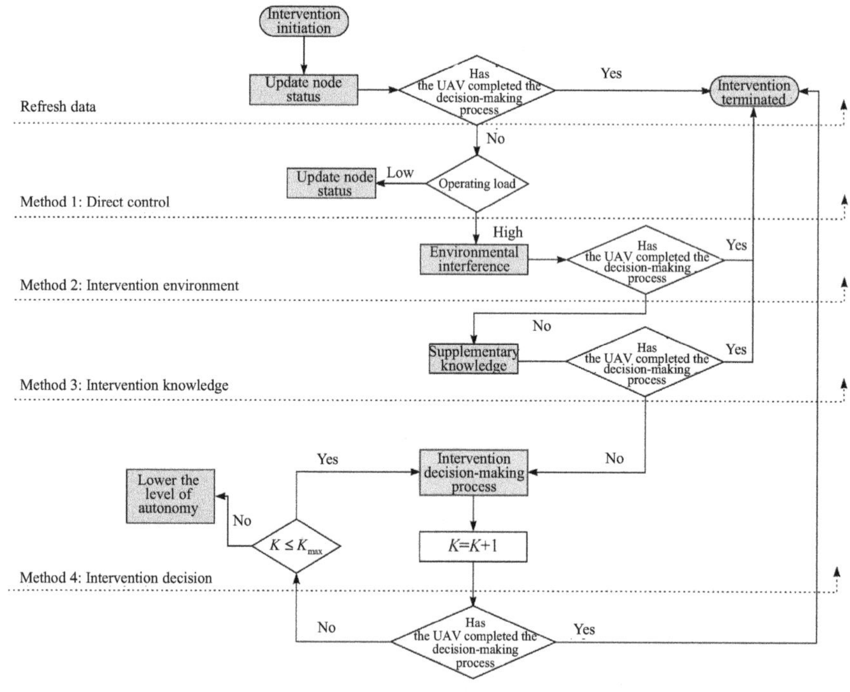

FIGURE 2.5 Limited-intervention engagement methods for manned aircraft

earlier prove ineffective, the manned aircraft can intervene in the autonomous decision-making process of the UAV by adjusting parameters such as narrowing the range of available actions or restricting the execution of complex maneuvers. Adjustments exceeding certain thresholds $(K > K_{max})$ will reduce the autonomy level of the UAV, thereby constraining its autonomous capabilities and necessitating greater operator involvement in control functions. The implementation of these intervention methods depends on actual battlefield conditions. The objective is to maximize the utilization of the UAV's autonomous intelligence while minimizing operator mental workload. When optimal decision outcomes cannot be guaranteed, these methods seek to achieve suboptimal or noninferior solutions.

The limited-intervention mechanism of manned aircraft embodies the hierarchical nature of decision-making requirements and the limited nature of intervention in the UAV's autonomous decision-making, ensuring dynamic adaptation capability for collaborative decision-making between manned aircraft and UAVs in actual battlefield environments.

2.5 AUTONOMY LEVEL ADJUSTMENT MECHANISM FOR UAV

In certain scenarios, allowing systems to make decisions autonomously may lead to catastrophic consequences. This necessitates human intervention when required, giving rise to the concept of variable autonomy [56–58]. The hierarchical control framework for autonomous robots, developed in the 1980s, employs independent, parallel modules that handle specific events. Each module interacts directly with its environment while maintaining minimal communication with other components. The primary characteristic of this framework is its event-stimulated reactive response and state transition mechanism [59]. Researchers at the National Aeronautics and Space Administration (NASA) employed variable autonomy models to achieve human-machine collaboration, developing hybrid-initiative variable autonomy technology aimed at simulating human-robot cooperation and interoperability in actual space environments. Autonomy level adjustment, as a process, depends on the type of actions the UAV executes, which in turn is determined by currently occurring events. Therefore, autonomy level adjustment can be defined as the process of modifying relevant parameters in the decision-making process based on current event types, thereby altering decision outcomes and mission execution results.

2.5.1 Autonomy Level Evaluation Methods

Selecting a set of metrics that accurately characterizes autonomous performance is crucial for evaluating autonomy levels and remains a challenging problem in the field of variable autonomy. Within the context of limited intervention-based collaborative decision-making between manned aircraft and UAVs, the most critical issue in variable autonomy is the intervention by the manned aircraft in the decision-making process of the UAV. During this process, intervention from the manned aircraft is required in two scenarios: when there is insufficient knowledge for decision-making, or when there is a lack of authority to execute actions. Therefore, "knowledge" and "authority" can be designated as the primary parameters for evaluating the autonomy level of UAVs.

In this case, a UAV A_i can execute the decision-making process d, denoted as Aq_i, only if the following two conditions are satisfied:

(1) UAV A_i possesses the knowledge $k\ (\text{know} == \text{true})$ required for decision-making.

(2) UAV A_i possesses the authority $c\ (\text{can} == \text{true})$ required to execute the decision.

This definition can be formally expressed as follows:

$$\text{If}\,(A_i\ \text{know} == \text{true})\,\text{and}\,(A_i\ \text{can} == \text{true}) \rightarrow A_i : \mathrm{Aq}_i \tag{2–1}$$

The autonomous decision-making process of UAVs is divided into four phases, that is, perception-reasoning-planning-execution. The evaluation criteria for autonomy differ across these phases. Autonomy in the perception and reasoning phases is primarily manifested at the knowledge dimension, while the planning phase requires integrated consideration of both knowledge and authority. The execution phase, conversely, is mainly reflected at the authority dimension. Should autonomy deficiencies be identified during any operational phase, intervention from the manned aircraft becomes necessary.

Let a_i denote the autonomy level of UAV A_i, $0 \le a_i \le 1$.

If $a_i \approx 0$, then a UAV A_i possesses no autonomy, whereas if $a_i \approx 1$, it denotes that a UAV A_i is fully autonomous.

For an event e, the knowledge possessed by a UAV A_i is k_e, expressed by the function:

$$\text{know}(e) = k_e, \quad 0 \le k_e \le 1 \tag{2–2}$$

The authority possessed by a UAV A_i to handle event e is c_e, expressed by the function:

$$\text{can}(e) = c_e, \quad 0 \le c_e \le 1 \tag{2–3}$$

Definition 2.1: The autonomy level of UAV A_i for event e in the perception and reasoning phases is determined by its knowledge proficiency regarding the event:

$$\forall e \exists A_i : \text{know}(e) \Rightarrow a_i = k_e \cdot \gamma \tag{2–4}$$

where γ is the intervention coefficient, indicating whether the manned aircraft intervenes. The value of γ depends on the state and performance of the UAV. $\gamma = 0$ indicates no intervention from the manned aircraft, while $\gamma = 1$ indicates intervention.

Definition 2.2: The autonomy level of UAV A_i for event e in the mission planning phase depends on both its knowledge and authority levels regarding the event:

$$\forall e \exists A_i : \text{know}(e) \wedge \text{can}(e) \Rightarrow a_i = (\sigma_1 k_e + \sigma_2 c_e) \cdot \gamma \Leftrightarrow \min[k_e, c_e] \ge T \tag{2–5}$$

where T denotes the minimum autonomy threshold, σ_1 σ_2 are coefficients with a value range of [0, 1], and min $[k_e, c_e] \ge T$ indicates that the UAV possesses autonomy in this phase when k_e and c_e exceed the autonomy threshold.

Definition 2.3: The autonomy level of UAV A_i for event e in the mission execution phase depends on its authority level regarding the event:

$$\forall e \exists A_i : \text{can}(e) \Rightarrow a_i = c_e \cdot \gamma \tag{2–6}$$

As shown in Figure 2.6, assuming both knowledge and authority are linear functions of the decision-making process, the triangular region T bounded by authority and knowledge represents the autonomy threshold. By adjusting the slopes of the linear functions for authority and knowledge, one can modulate the range of the autonomy threshold. The knowledge and authority levels of the UAV must exceed T to meet the autonomy

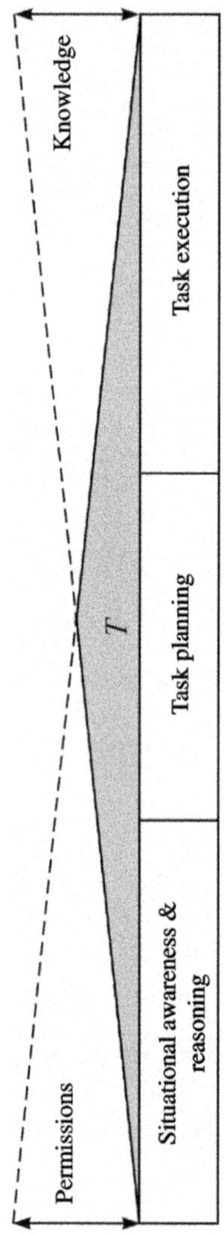

FIGURE 2.6 The relationship between UAV autonomy level evaluation and mission phases

requirements for each phase. The mission planning phase imposes the highest demands on knowledge and authority levels, whereas the SA and reasoning phases and the mission execution phase require sufficient knowledge and authority.

According to the definition of UAV autonomy levels in Table 2.1, the knowledge and authority requirements for each autonomy level during mission operations are illustrated in Figure 2.7. The gray-filled boxes indicate satisfied conditions (e.g., "know"). The white-filled boxes indicate unsatisfied conditions (e.g., "can"). The first column on the left denotes the autonomy level, while the last column on the right represents the mission output of the UAV platform at that respective autonomy level. A UAV at A_1 (instruction control) possesses no level of knowledge or authority, and it can only utilize its own sensor payload to output situational parameters. A UAV at A_2 (operator decision) possesses only the knowledge required for reasoning and the knowledge required for planning. At this autonomy level, the UAV can provide SA results and mission planning parameters to the manned aircraft, but it cannot make any decisions. The action selection for the UAV is performed by the manned aircraft operator. A UAV at A_3 (consent management) possesses the knowledge required for reasoning and planning, as well as the authority required for mission planning, and it can autonomously select actions. At this level, the UAV can provide mission planning results to the manned aircraft and execute them after obtaining consent from the manned aircraft operator. A UAV at A_4 (exception management) can independently complete the entire mission decision-making process and possesses the authority for autonomous execution.

2.5.2 Dimensions of Autonomous Adjustment

Referring to the French Ground Robotics Project, the variable autonomy model is designed with two dimensions: the descriptive dimension and the normative dimension [44]. The descriptive dimension refers to the actions an agent is capable of performing, describing the agent's self-sufficiency. The normative dimension refers to the actions an agent is permitted to perform under certain constraints, describing the degree of external constraints imposed on the agent [60]. Based on the action sets defined for each autonomy level, horizontal adjustment of the autonomy level is achieved by restricting the scope of the action set within the same level, while vertical adjustment of the autonomy level is realized through switching between different autonomy levels.

Task process →

	Perception	Reasoning	Planning		Execution	Task output
A_4		know	know	can	can	Execution result
A_3		know	know	can	~can	Task result
A_2		know	know	~can	~can	Cognitive result
A_1		~know	~know	~can	~can	Situation parameters
		Cognitive knowledge	Planning knowledge	Planning authority	Execution permission	

FIGURE 2.7 The requirements for knowledge and authority at each autonomy level

Figure 2.8 illustrates the two adjustment dimensions of UAV autonomy levels. Each level is divided into two layers: the upper layer represents the set of commands or actions available, while the lower layer indicates the proportion of decision-making authority allocated between the manned aircraft operator and the UAV during task execution. At lower autonomy levels, the UAV's actions are almost entirely controlled by the operator, whereas at higher autonomy levels, the operator's authority accounts for a smaller portion, and the UAV's actions are nearly autonomous. The horizontal adjustment method regulates the range of actions the UAV performs while keeping its autonomy level fixed. The vertical adjustment method switches the UAV's autonomy level, altering the interaction method for collaborative decision-making between the manned aircraft and the UAV. These two-dimensional autonomy adjustment methods allow for controlling the intensity of autonomy level adjustments based on the situational context. For example, horizontal adjustment alone may suffice for handling emergencies with lower danger levels. In situations of higher danger, the autonomy level should be reduced to allow greater operator involvement, thereby better addressing the threat.

The actions selected by the UAV through mission planning belong to the descriptive dimension, while the actions executed by the UAV under authority constraints belong to the normative dimension. Generally, operational complexity correlates with the required authorization levels:

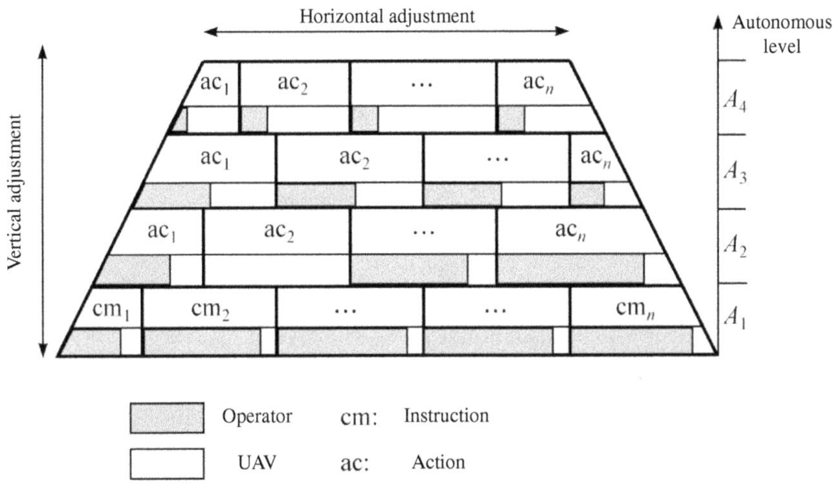

FIGURE 2.8 The two adjustment dimensions of UAV autonomy levels

Higher-complexity actions like weapon deployment demand elevated authorization tiers, while simpler actions such as returning to base require lower-level approval. This authorization hierarchy reflects the mission impact potential, with higher-complexity actions typically yielding greater operational effectiveness, such as successful target engagement. Therefore, during mission planning, the UAV tends to select actions of higher complexity in the descriptive dimension. At this point, the manned aircraft operator can adjust the normative dimension based on the actual situation to constrain certain actions of the UAV. For example, if the target threat is low, the constraints of the normative dimension will be less pronounced, and the UAV can execute more complex actions. When the target threat potential is high, the normative dimension will narrow the range of executable actions for the UAV, leading it to choose safer actions. Consequently, it is necessary to classify UAV autonomy levels according to action complexity.

It is assumed that the mission actions of a UAV can be divided into three levels. A_D is a complex action involving reasoning processes like attacking targets. A_P represents actions generated by fixed conditions like circling due to communication delays. A_N is the basic action of a UAV, with the lowest complexity such as turning, returning, and reactive avoidance. Action complexity is classified into high-level actions, mid-level actions, and low-level actions, as shown in Figure 2.9.

Definition 2.4: Let actions denote the set of actions: $\text{Actions} = \{A_D \cup A_P \cup A_N\}$. A_H denotes the set of high-level actions, $A_D \to A_H$. A_I denotes the set of mid-level actions, $A_P \to A_I$. A_L denotes the set of low-level actions, $A_N \to A_L$.

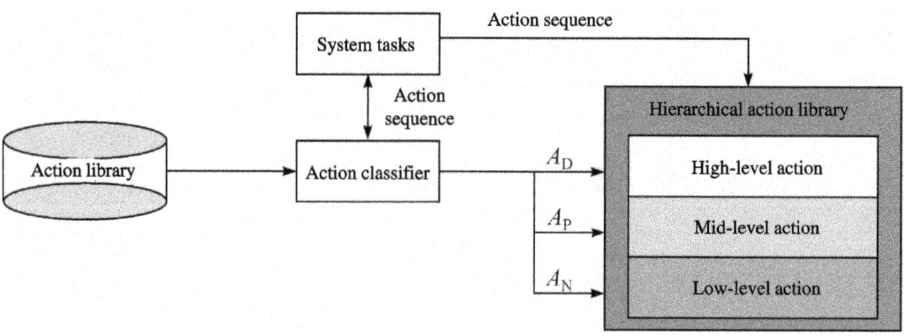

FIGURE 2.9 The classification process of action complexity

Definition 2.5: Let $L = \{A_1, A_2, A_3\}$ denote the autonomy level for action assignment. The action libraries of A_4 and A_3 are identical, differing only in execution authority. Hence, they do not appear in L. The action hierarchy assignment function Ld(a) for each autonomy level is defined as follows:

$$\mathrm{Ld}(a) = \begin{cases} a \in A_3, & a \in (A_\mathrm{H} \cup A_\mathrm{I} \cup A_\mathrm{L}) \\ a \in A_2, & a \in (A_\mathrm{I} \cup A_\mathrm{L}) \\ a \in A_1, & a \in A_\mathrm{L} \end{cases} \tag{2–7}$$

Equation (2–7) represents the action hierarchies contained at each autonomy level. Higher autonomy levels encompass actions from all three hierarchies, while lower autonomy levels contain only mid-level or low-level actions, as illustrated in Figure 2.10.

Definition 2.6: As shown in Figure 2.7, vertical adjustment of autonomy levels refers to the matching and switching between levels based on the UAV's knowledge and authority levels during different mission phases and the conditions they satisfy.

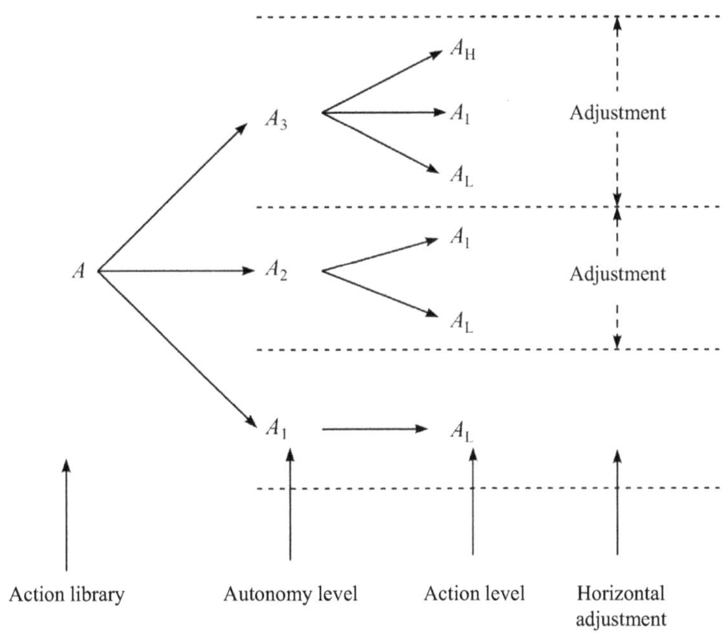

FIGURE 2.10 Relationship between action hierarchy and autonomy level

Definition 2.7: As illustrated in Figure 2.10, horizontal autonomy adjustment refers to modifying the scope of actionable tasks executable by a UAV within the same autonomy level.

Definition 2.8: The adjustment of UAV autonomy levels correlates directly with the intensity of intervention from the manned aircraft. Weaker operator intervention typically corresponds to horizontal adjustments within the same autonomy level, while stronger intervention involves vertical transitions between different autonomy tiers.

2.5.3 Variable Autonomy Model for UAV

The variable autonomy model for UAV partitions each autonomy level into action hierarchies within the traditional autonomy level structure. During mission operations, the UAV can be adjusted either among different autonomy levels or across different hierarchies within the same autonomy level. Among studies on the evaluation of UAV autonomy levels, the hierarchical autonomy structure, and adjustment methods, the decision-making process know(event) $\wedge$ can(event) $\rightarrow$ decide(event) can be realized, as shown in Figure 2.11.

The variable autonomy model consists of the event analysis module (EAM), the situational awareness module (SAM), the autonomy level management module (ALMM), and the mission effectiveness evaluation module (MEEM). The EAM and SAM are responsible for processing event information to form SA and assisting the UAV in situational reasoning tasks. The ALMM is responsible for evaluating and adjusting

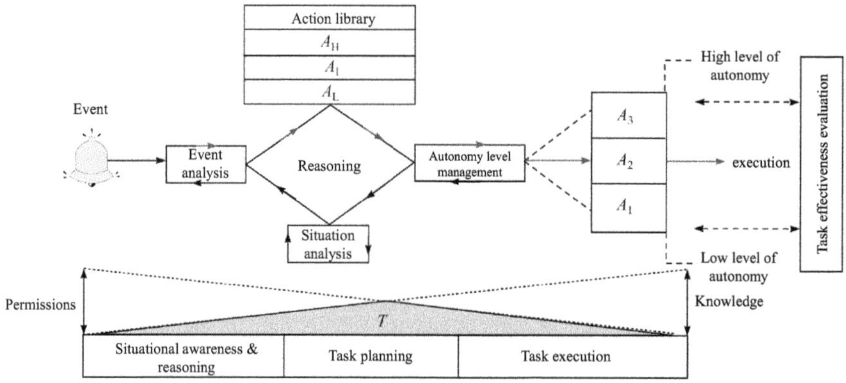

FIGURE 2.11 Variable autonomy model for UAV

the autonomy level of the UAV. The MEEM assesses the execution process and results of the mission, providing a reference for subsequent mission operations.

In dynamic environments, as the system state changes, the required values for knowledge and authority levels at each autonomy level will change constantly. This requires the system to adjust autonomy level-related parameters based on its own state to ensure the UAV operates at an appropriate autonomy level.

2.6 LIMITED-INTERVENTION COLLABORATIVE DECISION-MAKING AND OPERATIONAL SCHEDULING MECHANISM FOR MANNED AIRCRAFT – UAV

During the operation of the collaborative decision-making system, the manned aircraft operator assumes five task roles, including planning, guiding, monitoring, intervening, and learning, as illustrated and interpreted in Figure 2.12.

(1) Planning the mission before the UAV's autonomy is activated.

(2) Guiding the UAV to complete the mission.

(3) Monitoring the process by which the UAV accomplishes the mission.

(4) Intervening to ensure the UAV completes the mission normally.

(5) Learning from the UAV's mission execution process to serve as a reference for future planning.

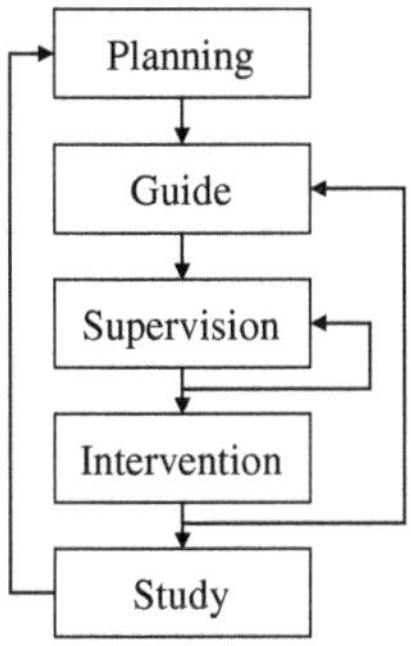

FIGURE 2.12 Limited-intervention operational loop for manned aircraft

The demand for the planning and guiding stages increases with the complexity of the mission, while monitoring and intervention depend on the effectiveness of planning and guiding.

The operational scheduling model framework for the limited-intervention collaborative decision-making system between manned aircraft and UAV can be divided into the system monitoring module (SMM) and the system control module (SCM). As shown in Figure 2.13, the SCM consists of the manned aircraft intervention module, the UAV variable autonomy module, and the UAV autonomous decision-making module, which collectively fulfill the primary functions of the supervisory control system. Among these, the manned aircraft intervention module facilitates interaction between the manned aircraft and UAV and implements intervention in the UAV variable autonomy module. Simultaneously, the UAV variable autonomy module can control the decision-making structure and form of the UAV, thereby influencing the decision outcomes. The SMM is responsible for monitoring the status of the manned aircraft operator, the status of the UAV, and assessing the environmental threat level to enable dynamic adjustments of relevant parameters in the control structure. For example, when the operator is under high stress, the autonomy level tends

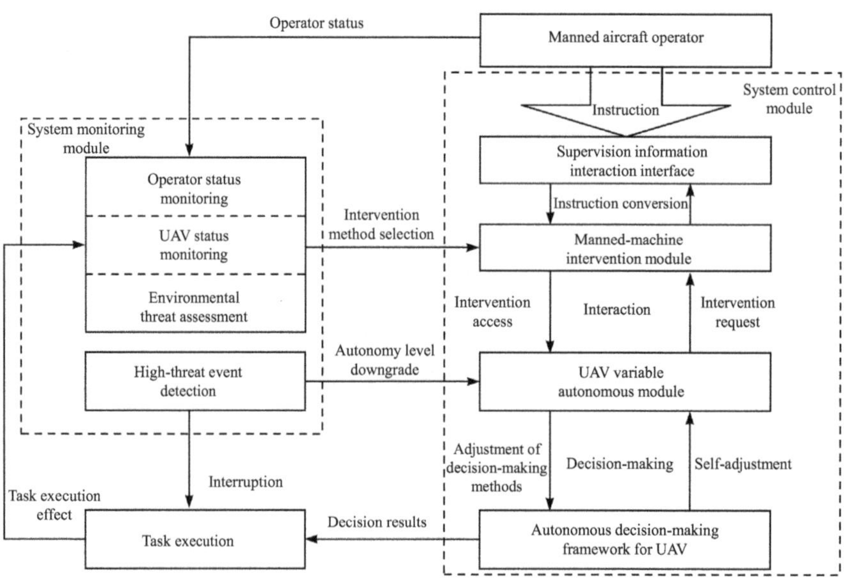

FIGURE 2.13 Operational scheduling model framework for the limited-intervention collaborative decision-making system between manned aircraft and UAV

to be increased. At the same autonomy level, for high-threat missions, the UAV tends to execute tasks with lower complexity. When the UAV's task completion level is high, the current autonomy level is maintained, whereas when the task completion level is low, the autonomy level range is appropriately reduced. Additionally, high-threat events are prioritized for handling, primarily by reducing the autonomy level to allow greater human involvement in task execution, or even interrupting the task execution process to ensure the safety of the UAV.

2.7 CHAPTER SUMMARY

This chapter analyzes the limited-intervention collaborative decision-making mechanism between manned aircraft and UAVs based on human-machine supervisory control technology. The objective is to prevent the manned aircraft operator from intervening in the underlying control systems of the UAV, thereby allocating cognitive resources to higher-level tasks. Hence, it is necessary to adopt the autonomous control and decision-making capabilities of UAVs and design reasonable intervention mechanisms to better achieve human-machine collaborative decision-making.

The main work accomplished in this chapter includes the following key points:

(1) Defined the concepts of limited intervention-based autonomous decision-making and autonomy levels for UAV.

(2) Designed a manned aircraft limited-intervention mechanism incorporating intervention trigger strategies and intervention methods.

(3) Defined the autonomous adjustment dimensions for UAVs and established a variable autonomy level adjustment model.

(4) Established an operational scheduling model for limited-intervention collaborative decision-making between manned aircraft and UAVs.

CHAPTER 3

Limited-Intervention Collaborative Decision-Making Model between Manned Aircraft and UAVs

BASED ON THE DESCRIPTION of the limited-intervention collaborative decision-making tasks between manned aircraft and UAVs, this chapter establishes models for collaborative obstacle avoidance decision-making, threat avoidance decision-making, and attack decision-making. Through case simulations, the collaborative decision-making models are analyzed and discussed here.

3.1 DEFINITION OF LIMITED-INTERVENTION COLLABORATIVE DECISION-MAKING TASKS BETWEEN MANNED AIRCRAFT AND UAV

To investigate the modeling method for limited-intervention collaborative decision-making between manned aircraft and UAVs, the collaborative decision-making task is described as a manned aircraft and a UAV collaboratively executing a mission to attack ground targets. During the mission, the UAV must avoid obstacles and sudden threats. After entering the target

DOI: 10.1201/9781003791713-3

area, it needs to make attack decisions against ground targets. Throughout this process, the manned aircraft provides limited intervention to accomplish the collaborative decision-making task, as shown in Figure 3.1.

In collaborative decision-making tasks, the UAV can provide the manned aircraft with situational parameters, SA results, mission planning parameters, mission planning results, and perform weapon-delivery tasks. The manned aircraft, on the other hand, acts as a supervisor and manager, monitoring and intervening in the mission-execution process, controlling the UAV to conduct strikes, and, when necessary, executing strike missions against ground targets itself. Due to the presence of unknown and

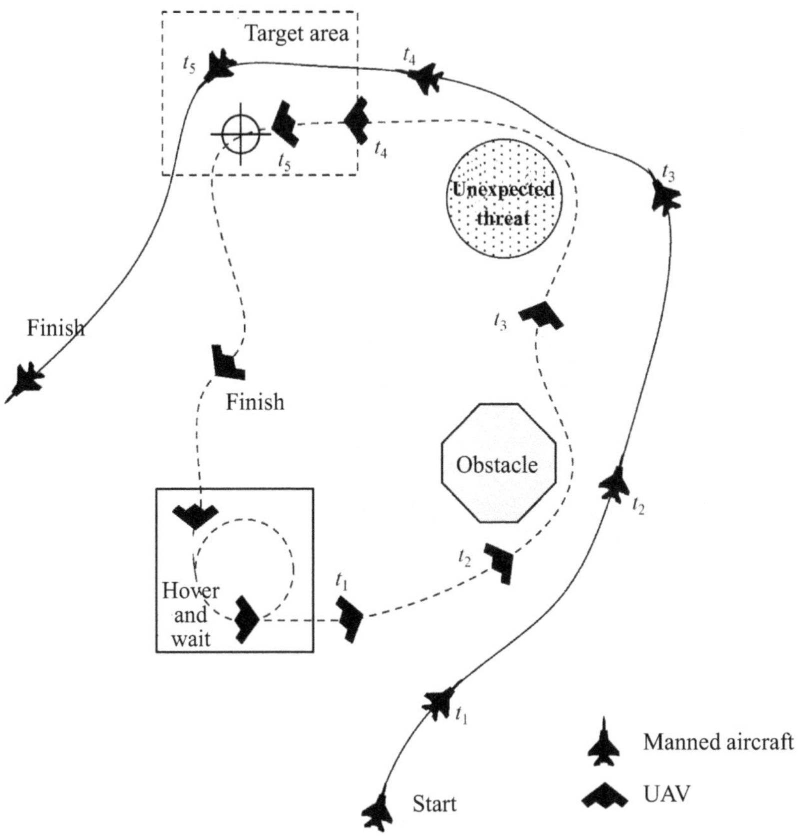

FIGURE 3.1 Description of the limited-intervention collaborative decision-making tasks between manned aircraft and UAV

dynamic environmental factors, both the manned aircraft and UAVs face varying levels of threats. Since the manned aircraft has its own independent mission objectives, the operator does not have sufficient bandwidth or time to control the low-level actions of the UAV such as turning, accelerating, or decelerating. To reflect the limited-intervention characteristic of collaborative decision-making between manned aircraft and UAV, the focus is on modeling and analyzing three key collaborative decision-making aspects during the mission: obstacle avoidance, sudden threat avoidance, and ground attack.

Some researchers have compared the mission effectiveness of systems under the consent management mode and the exception management mode, and they found that systems under the consent management mode achieve the highest mission effectiveness and the highest UAV survival rate. Under the consent management mode, operators demonstrate the highest level of SA of the battlefield, followed by manual operation, whereas the exception management mode results in the lowest level of operator SA. Therefore, in the modeling, it is assumed that the initial autonomy level of the UAV is set to the consent management mode. When necessary, the manned aircraft provides limited intervention in the decision-making process of the UAV, and the autonomy level of the UAV is adjusted accordingly.

3.2 COLLABORATIVE OBSTACLE AVOIDANCE DECISION-MAKING MODEL FOR MANNED AIRCRAFT AND UAV

The obstacle-avoidance decision-making process for UAVs includes (1) detecting obstacles and triggering the obstacle-avoidance task, (2) acquiring information related to obstacles, (3) autonomously reasoning to select an obstacle-avoidance solution, and (4) executing obstacle-avoidance actions under the manned aircraft's limited-intervention decision-making.

3.2.1 Action Hierarchy Classification for UAV Obstacle Avoidance

Assuming the initial autonomy level of the UAV is set to the consent-management mode, the optional action library for the UAV during obstacle avoidance is shown in Table 3.1.

As shown in Table 3.1, when the UAV encounters an obstacle, it has four optional actions depending on the obstacle type and environmental factors. When the obstacle type is known, the UAV can directly select autonomous obstacle avoidance. When the obstacle type is unknown, the UAV transmits the obstacle image to the manned aircraft and chooses either

TABLE 3.1 Obstacle Avoidance Action Library for UAV

Action Number	Action Name	Selection Criteria	Action Hierarchy
A_1	Obstacle avoidance	Known obstacles	High
A_2	Deceleration wait instructions	Unknown obstacles, long distance	Middle
A_3	Hovering and waiting for instructions	Unknown obstacles, medium distance	Middle
A_4	Return	Communication interruption	High

TABLE 3.2 Environmental Factors Considered in UAV Obstacle Avoidance

Environmental Factor Label	Environmental Factor Name
V_1	Known obstacles
V_2	Communication interruption
V_3	Close obstacle distance
V_4	Unknown obstacles
V_5	Far obstacle distance

to hover or reduce speed based on the obstacle distance while awaiting control commands from the manned aircraft. The manned aircraft operator identifies the obstacle type and controls the UAV to perform obstacle avoidance. When communication is interrupted, the UAV selects to return to the departure point.

3.2.2 FCM Model for UAV Obstacle Avoidance Decision-Making

This subsection utilizes the description of on/off states of nodes in the binary FCM model to represent the UAV's action selection. The environmental factors that need to be considered during obstacle-avoidance decision-making by the UAV are shown in Table 3.2.

The FCM model for UAV obstacle avoidance decision-making is shown in Figure 3.2, with node definitions provided in Table 3.3. In the figure, the symbol "+" indicates a positive influence between nodes, while "–" indicates a negative influence between nodes. The polarity of the influence relationships describes the trade-offs among various nodal factors and their interactions during the decision-making reasoning process. The value range of each node is {0,1}, where "1" indicates that the node state is active (i.e., the action is selected or the environmental factor exists), and

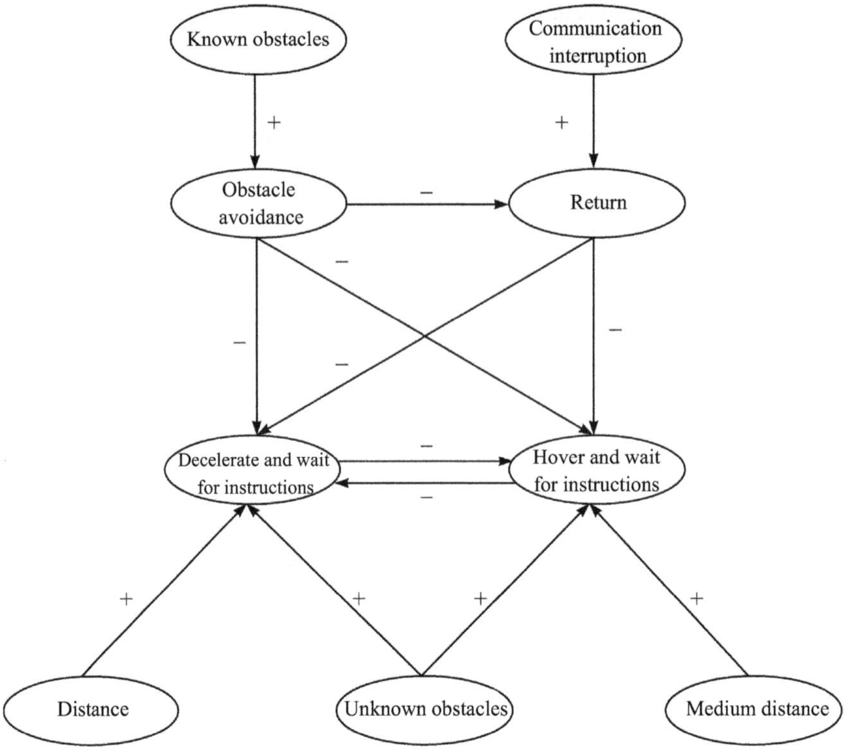

FIGURE 3.2 FCM model for UAV obstacle avoidance decision-making

TABLE 3.3 Node Definitions for the FCM Model of UAV Obstacle Avoidance Decision-Making

Node Label	Node Name
A_1	Obstacle avoidance
A_2	Deceleration wait instructions
A_3	Hovering and waiting for instructions
A_4	Return
V_1	Known obstacles
V_2	Communication interruption
V_3	Close obstacle distance
V_4	Unknown obstacles
V_5	Far obstacle distance

"0" indicates that the node state is inactive. The reasoning process of the FCM model for UAV obstacle avoidance decision-making converges into two possible states, a fixed point or a limit cycle. If the FCM converges into a fixed point, it indicates a definitive action selection, whereas when it converges to a limit cycle, it signifies that the UAV has multiple actionable options in the given situation, requiring intervention from the manned aircraft.

The node state transition function is a binary step function:

$$f(x)=\begin{cases}0, & x\leq 0\\ 1, & x>0\end{cases} \tag{3–1}$$

The iteration formula is as follows:

$$C(k+1)=f\left[C(k)W\right] \tag{3–2}$$

The state vector is as follows:

$$\vec{C}=[A_1,A_2,A_3,A_4\vdots V_1,V_2,V_3,V_4,V_5] \tag{3–3}$$

The adjacency weight matrix of the FCM model for the UAV obstacle avoidance decision-making is shown in Table 3.4.

3.2.3 Limited-Intervention Decision Model for Obstacle Avoidance in Manned Aircraft

Upon completing the obstacle-avoidance decision-making process, the UAV adjusts its autonomy level according to the actions selected through its reasoning. Variations in autonomy levels will lead to differences in the interactive methods between the UAV and the manned aircraft, ultimately determining the UAV's mission execution approach.

TABLE 3.4 Adjacency Weight Matrix of the FCM for UAV Obstacle Avoidance Decision-Making

Node	A_1	A_2	A_3	A_4	V_1	V_2	V_3	V_4	V_5
A_1	0	−1	−1	−1	0	0	0	0	0
A_2	0	0	−1	0	0	0	0	0	0
A_3	0	−1	0	0	0	0	0	0	0
A_4	0	−1	−1	0	0	0	0	0	0
V_1	1	0	0	0	0	0	0	0	0
V_2	0	0	0	1	0	0	0	0	0
V_3	0	0	1	0	0	0	0	0	0
V_4	0	1	1	0	0	0	0	0	0
V_5	0	1	0	0	0	0	0	0	0

Example 1:

The UAV encounters an unknown obstacle in close proximity and a communication interruption. The initial node state vector is $\vec{C} = [0,0,0,0 \vdots 0,1,1,1,0]$, and the iterative computation is performed until a stable point is reached, and the output result is $\vec{C} = [0,0,1,1 \vdots 0,1,1,1,0]$. The results indicate that the reasoning outcome for the UAV in this case is to select both hovering while waiting for instructions from the manned aircraft and returning. This means that under the consent management mode, the UAV will perform the hovering action within a given time limit while waiting for intervention instructions from the manned aircraft. If no feedback is received beyond the time limit, it will reduce the mission hierarchy level and execute the return action.

In UAV obstacle-avoidance decision-making tasks, if an unknown obstacle is encountered, the UAV actively requests intervention from the manned aircraft due to its lack of information regarding the unknown obstacle. Therefore, the intervention method of the manned aircraft should primarily focus on knowledge-level intervention, supplementing the UAV's information about the obstacle to assist in completing obstacle avoidance. The limited-intervention decision-making model for the manned aircraft in this scenario is shown in Figure 3.3.

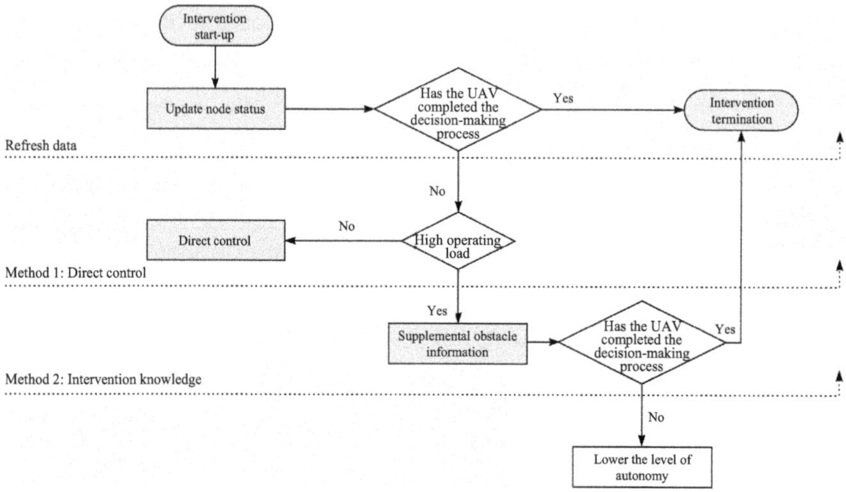

FIGURE 3.3 Limited-intervention decision model for obstacle avoidance in manned aircraft

In scenarios requiring manned aircraft intervention, the UAV's autonomy undergoes horizontal adjustment within the same level due to insufficient information about unknown obstacles. Since the autonomy level remains unchanged, the interaction mode between the UAV and the manned aircraft stays consistent. However, this horizontal adjustment narrows the scope of actionable options available to the UAV's decision-making system.

3.3 COLLABORATIVE THREAT AVOIDANCE DECISION-MAKING MODEL FOR MANNED AIRCRAFT AND UAV

The threat avoidance process of the UAV can be divided into four phases, including (1) triggering of sudden threat events, (2) acquisition of threat information, including threat value, threat distance, and whether it is a known threat, (3) decision-making reasoning based on threat information to select threat avoidance actions, and (4) adjustment and execution of threat avoidance actions under the manned aircraft's limited-intervention decision-making.

3.3.1 Action Hierarchy Classification for UAV Threat Avoidance

When UAV encounters sudden threats, it has six actionable options: (1) If the threat value is high and the threat source is known, the UAV can either autonomously attack or transmit situational information to the manned aircraft to request the manned aircraft to attack the target. (2) If the threat value is low, it chooses to avoid the threat. (3) If the threat source is unknown, and the UAV lacks knowledge about this threat in its knowledge base, it selects either hovering or speed reduction based on the threat distance while awaiting instructions from the manned aircraft. (4) If the sudden threat is too close, leaving no time for temporary threat avoidance planning, it may execute reactive avoidance, such as turning with the minimum turning radius. (5) If communication is interrupted during the mission, it conservatively chooses to return. (6) The actionable library for the UAV during threat avoidance is shown in Table 3.5.

3.3.2 Autonomous Decision-Making FCM Model for UAV Threat Avoidance

When a UAV encounters a sudden threat, its response depends on threat identification. If the threat matches a known type in its database, the UAV can autonomously execute avoidance reasoning. If the threat is unknown,

TABLE 3.5 Threat Avoidance Action Library for UAV

Action Number	Action Name	Selection Criteria	Action Level
A_1	Attack decision	Known threats, greater value	High
A_2	Avoidance	Known threats, low value	High
A_3	Deceleration, wait instructions	Unknown threats, long distance	Middle
A_4	Hovering and waiting for instructions	Unknown threats, medium distance	Middle
A_5	Reactive avoidance	Unknown threats, short distance	Low
A_6	Return	Communication interruption	Low

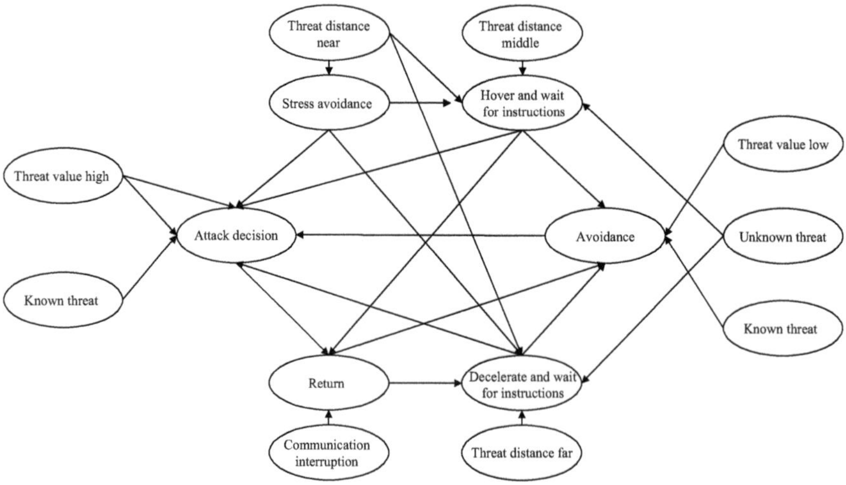

FIGURE 3.4 FCM model for UAV threat avoidance decision-making

the UAV transmits the threat data to the manned aircraft operator, who then provides limited intervention to guide the avoidance decision-making process. The FCM model for UAV threat avoidance decision-making is shown in Figure 3.4, with node definitions provided in Table 3.6 and the adjacency weight matrix shown in Table 3.7.

Employing a binary FCM, its node state transition function and iteration formula are as shown in Equations (3–1) and (3–2), respectively.

The reasoning process of the FCM for UAV threat avoidance decision-making converges into two possible states, including a fixed point or a

TABLE 3.6 Node Definitions for the FCM of UAV Threat Avoidance Decision-Making

Node Label	Node Name
A_1	Attack decision
A_2	Avoidance
A_3	Deceleration wait instructions
A_4	Hovering and waiting for instructions
A_5	Reactive avoidance
A_6	Return
V_1	Close threat distance
V_2	Medium threat distance
V_3	Unknown threats
V_4	Far threat distance
V_5	Communication interruption
V_6	Known threats
V_7	Threat value

TABLE 3.7 Adjacency Weight Matrix of the FCM for UAV Threat Avoidance Decision-Making

Node	A_1	A_2	A_3	A_4	A_5	A_6	V_1	V_2	V_3	V_4	V_5	V_6	V_7
A_1	0	0	−1	0	0	0	0	0	0	0	0	0	0
A_2	−1	0	0	0	0	0	0	0	0	0	0	0	0
A_3	−1	−1	0	−1	0	0	0	0	0	0	0	0	0
A_4	−1	−1	−1	0	−1	0	0	0	0	0	0	0	0
A_5	−1	0	−1	−1	0	0	0	0	0	0	0	0	0
A_6	−1	−1	−1	−1	0	0	0	0	0	0	0	0	0
V_1	0	0	−1	−1	1	0	0	0	0	0	0	0	0
V_2	0	0	0	1	0	0	0	0	0	0	0	0	0
V_3	0	0	1	1	0	0	0	0	0	0	0	0	0
V_4	0	0	1	0	0	0	0	0	0	0	0	0	0
V_5	0	0	0	0	0	1	0	0	0	0	0	0	0
V_6	1	1	0	0	0	−1	0	0	0	0	0	0	0
V_7	1	1	0	0	0	0	0	0	0	0	0	0	0

limit cycle. If the FCM converges to a fixed point, it indicates a definitive action selection, whereas if it converges to a limit cycle, it signifies that the UAV has multiple actionable options under the given conditions, requiring intervention from the manned aircraft.

3.3.3 Limited-Intervention Decision Model for Threat Avoidance in Manned Aircraft

After the threat avoidance decision-making reasoning process is completed, the UAV will adjust its autonomy level based on the actions selected according to the reasoning results. Differences in autonomy levels will lead to variations in the interaction methods between the UAV and the manned aircraft, ultimately determining the mission execution approach of the UAV.

Example 2:

The UAV encounters a sudden threat during the mission, with the threat source known, a high threat value, and a close threat distance. The initial node state vector is $\vec{C} = [0,0,0,0,0,0 \vdots 1,0,0,0,0,1,1]$. Iterative computation is performed until the convergence point is reached, and the output node state vector is $\vec{C} = [0,1,0,0,0,1 \vdots 0,1,0,0,0,1,1]$. The results indicate that the actions selected by the UAV in this case are avoidance or return. Because the initial setting is the consent management mode, which tends to select higher-level actions, the UAV chooses threat avoidance.

Example 3:

When the threat distance changes from close to far while other conditions remain unchanged, the autonomous decision-making output of the UAV results in either an attack decision or an avoidance decision. This demonstrates that the threat avoidance decision-making FCM model considers the influence of threat distance on the selection between avoidance and attack actions. Under the conditions of this example, the UAV tends to select the attack decision as the intended action.

During the UAV's threat avoidance decision-making process, there are two scenarios that require intervention from the manned aircraft: (1) When a known threat has high value, the UAV may request the manned aircraft to assist in attacking the threat source to eliminate the threat. This corresponds to the environmental intervention decision approach by the manned aircraft. (2) When facing an unknown threat, the UAV cannot execute autonomous avoidance and requires the manned aircraft to either supplement threat information or assume direct control for evasion. This scenario corresponds to the knowledge-based intervention approach by the operator.

Similar to obstacle avoidance, manned aircraft intervention during threat avoidance missions triggers a horizontal adjustment of the UAV's autonomy, altering its scope of actionable options within the same

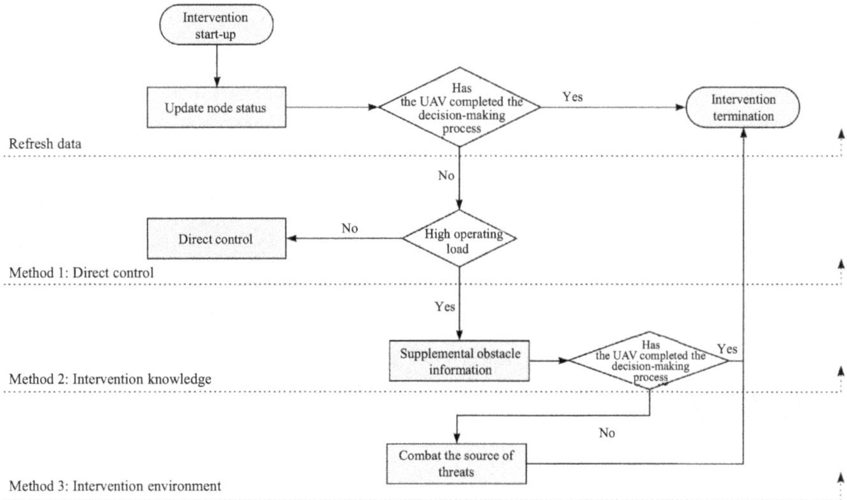

FIGURE 3.5 Limited-intervention decision model for threat avoidance in manned aircraft

operational tier. The autonomy level is reduced only when the UAV is unable to complete the mission at its current level. The limited-intervention decision-making model for threat avoidance is shown in Figure 3.5.

3.4 COLLABORATIVE ATTACK DECISION-MAKING MODEL FOR MANNED AIRCRAFT AND UAV

Compared to obstacle and threat avoidance, UAV attack decision-making involves greater complexity, encompassing target threat assessment, analysis of friendly force advantages, attack modeling, and other critical factors. Accordingly, intervention by the manned aircraft in UAV autonomous decision-making is more complex. As a result, the representational capacity of a binary FCM is insufficient to meet the demands of UAV attack decision-making. Here, extended FCM models such as Dynamic Fuzzy Cognitive Maps (DFCM) and Agent-based Fuzzy Cognitive Maps (ABFCM) are adopted for modeling.

The collaborative attack decision-making process can be described as follows: Based on the battlefield environment, the UAV performs autonomous decision-making. When autonomous decision-making becomes challenging, the UAV transmits intervention requests and information such as suggestions to deploy electronic countermeasures to the manned aircraft through auxiliary decision-making nodes. The manned aircraft then implements intervention for a UAV experiencing autonomous decision-making difficulties, either by deploying countermeasures to alter the battlefield environment or by adjusting parameters such as the structure and thresholds of the UAV's autonomous attack decision-making model, thereby collaboratively completing the attack decision.

The dynamic nature of the UAV's autonomous attack decision-making process is reflected as follows: the UAV receives a synthesized set of external environmental information $V = \{V_e, V_l, V_f\}$, where V_e represents external environmental influences, V_1 denotes influences from the manned aircraft platform, and V_f indicates influences from other platforms, and, simultaneously, it receives values from the concept node vector $\vec{C}$ and dynamically adjusts the FCM model.

The FCM modeling for UAV autonomous attack decision-making encompasses three aspects:

(1) Determination of auxiliary decision-making nodes
The auxiliary decision-making node serves as a bridge connecting the UAV and the manned aircraft in the model. It functions not only as an output node for the UAV's auxiliary decision-making results but also as an input node for the manned aircraft's collaborative decision-making.

(2) Design of the DFCM reasoning algorithm
FCM typically involves a concept node set C, an adjacency weight matrix W, an activation function f, and a node state value set $C(t)$ at time t. The mathematical representation is as follows:

$$c_i(t+1) = f\left(k_1 c_i(t) + k_2 \sum_{\substack{i=1 \\ j \neq i}}^{n} w_{ij} c_j(t) \right) \tag{3–4}$$

Here, nodes $c_i, c_j \in C$, weight $w_{ij} \in W$, node state values, $c_i(t)$ and $c_i(t+1)$ are the state values of node c_i at time t and $t+1$, respectively. $k_1, k_2 \in [0,1]$ are weight coefficients for the node state value and the comprehensive influence value of external nodes. Generally, $k_1 = 0$ or $k_1 = 1$, and $k_2 = 1$. The activation function f is used to constrain the reasoning results within a specific fuzzy interval, typically set to [0,1] or [−1,1].

The adjacency weight matrix of the model can be established through sample learning or based on expert experience. The weight matrix incorporates experts' prior knowledge and influences the decision-making reasoning outcomes. To avoid biased expert perspectives, the weight matrix of a typical FCM is generally constructed by integrating knowledge from multiple experts.

(3) Dynamic adjustment algorithm for the adjacency weight matrix
To address the problem of dynamic adjustment of the adjacency weight matrix W by the external environmental information set V, the specific adjustment formula is designed as follows:

$$w_{ij}(t+1) = f\{w_{ij}(t), V_k, a_k\} = \begin{cases} w_{ij}(t)(1+a_k V_k), & -1 \le w_{ij}(t+1) \le 1 \\ -1, & w_{ij}(t+1) < -1 \\ 1, & w_{ij}(t+1) > 1 \end{cases} \tag{3-5}$$

Here, V_k belongs to the external environmental information set V, representing an element within V (i.e., (V_e, V_l, V_f). a_k is the adjustment coefficient with a range of [−1, 1]. The adjacency weight is updated through a superposition of its previous state and the integrated influence exerted by external environmental information nodes. As w_{ij} changes during the iterative process, it is denoted as $w_{ij}(t)$ at time t. The adjustment of the adjacency weight is determined by V_k and a_k, where V_k determines the degree of dynamic influence from the external environment, and a_k is the magnitude and direction of the adjustment.

3.4.1 Attack Decision-Making Action Hierarchy Classification for UAV

As shown in Table 3.8, the UAV has four actionable options for attack decision-making: (1) When the relative advantage is significant, it selects

TABLE 3.8 Attack Decision-Making Action Library for UAV

Action Label	**Action Name**	**Selection Criteria**	**Action Level**
A_1	Attack	Relatively significant advantage	High
A_2	Quit	No attack advantage	High
A_3	Hovering and waiting for instructions	Difficulties in autonomous decision-making	Middle
A_4	Self-insured return	Communication interruption	Low

attack. (2) When the UAV has no attack advantage against the target, it chooses to withdraw. (3) When the UAV encounters difficulties in autonomous decision-making, it requests intervention from the manned aircraft and loiters while awaiting control instructions. (4) When communication is interrupted, the UAV selects self-preservation return.

3.4.2 FCM Model for UAV Autonomous Attack Decision-Making

3.4.2.1 Model Node Selection and Definition

(1) Terrain Complexity: The degree of terrain complexity typically affects combat operations and, ranked from high to low, includes mountains, plateaus, hills, basins, and plains. Higher terrain complexity corresponds to weaker attack advantage.

(2) Target Attack Difficulty: In actual combat scenarios, factors like target concealment, size, and velocity influence attack difficulty. Better concealment increases attack difficulty, while larger target size facilitates detection and attack. Higher target engagement difficulty corresponds to weaker attack advantage.

(3) Enemy Air Defense Intensity: Protecting military objectives is essential for defensive operations. More comprehensive air defense systems, higher modernization levels, and stronger coordination capabilities directly correspond to a reduction in the attack advantage and an increase in the enemy's threat level.

(4) Enemy Radar Status: Radar functions as the "eyes" of the modern warfare. Higher radar capability and modernization levels directly correlate with diminished friendly force attack advantage and an elevated enemy threat level.

(5) Target Distance: The relationship between distance advantage and target distance follows a unimodal peak function. When the target distance approaches the optimum value, distance advantage increases and corresponds to stronger attack advantages.

It is assumed that the minimum launch range of our missile is R_{min} and the maximum launch range is R_{max}. When the distance to the target aircraft satisfies $R > R_{max}$, the distance advantage is considered zero. As the relative distance decreases, the distance advantage gradually increases.

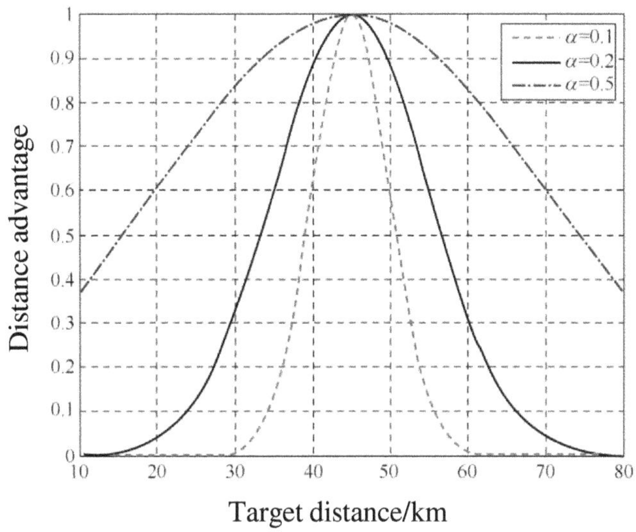

FIGURE 3.6 Distance advantage function curves under different values of α

When $R = 0.5(R_{max} + R_{min})$, the distance advantage is considered to reach its maximum value. With further reduction in relative distance, the distance advantage gradually decreases again.

Thus, a unimodal fuzzy function can be utilized to construct the distance advantage function via:

$$T = e^{-\left(\frac{R-R_0}{\sigma}\right)^2} \tag{3-6}$$

Here, $R_0 = \frac{1}{2}(R_{max} + R_{min})$, $\sigma = \alpha(R_{max} + R_{min})$, and α are the fuzzy factor. Figure 3.6 then demonstrates the distance advantage function curves under different values of α when $R_{max} = 80\text{km}$ and $R_{min} = 10\text{km}$.

3.4.2.2 Model Construction and Reasoning Algorithm Design

Figure 3.7 shows the FCM model for UAV autonomous attack decision-making, with node definitions provided in Table 3.9.

In Table 3.9, $C_3 \sim C_7$ are the input nodes, while C_8 is an auxiliary decision-making node responsible for reserving an interface for manned

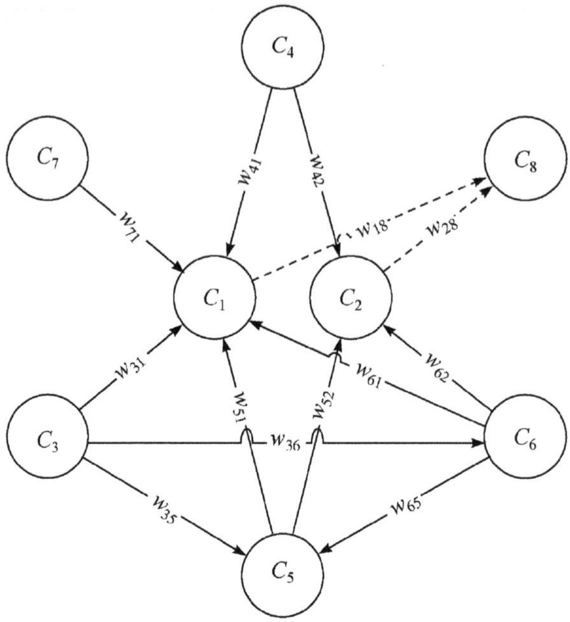

FIGURE 3.7 FCM model for UAV autonomous attack decision-making

TABLE 3.9 Node Definitions for the FCM Model of UAV Autonomous Attack Decision-Making

Node Label	Node Name
C_1	Our attack advantage
C_2	Target threat level
C_3	Terrain complexity
C_4	Target attack difficulty
C_5	Enemy air defense strength
C_6	Enemy radar status
C_7	Target distance
C_8	Required interference intensity

aircraft intervention assistance. $C_1 \sim C_2$ are the output nodes of the FCM model reasoning, outputting the attack advantage and the target threat level. Simultaneously, $C_1 \sim C_2$ serve as input nodes to C_8, used for reasoning of the auxiliary decision results. C_3, C_4, C_5, and C_6 have negative influences on C_1, while C_7 has a positive influence on C_1, C_4, C_5, and C_6 have positive influences on C_2.

The weight matrix of the model is as follows:

$$W = \begin{bmatrix} 0 & 0 & 0 & 0 & 0 & 0 & 0 & -0.1 \\ 0 & 0 & 0 & 0 & 0 & 0 & 0 & 0.7 \\ -0.65 & 0 & 0 & 0 & 0.1 & 0.2 & 0 & 0 \\ -0.35 & 0.2 & 0 & 0 & 0 & 0 & 0 & 0 \\ -0.8 & 0.85 & 0 & 0 & 0 & 0 & 0 & 0 \\ -0.7 & 0.6 & 0 & 0 & 0.55 & 0 & 0 & 0 \\ 0.5 & 0 & 0 & 0 & 0 & 0 & 0 & 0 \\ 0 & 0 & 0 & 0 & 0 & 0 & 0 & 0 \end{bmatrix}$$

The reasoning process of the FCM model for UAV autonomous attack decision-making is divided into two steps:

(1) The node states of $C_1 \sim C_2$ are derived through reasoning based on the model in Figure 3.7.

(2) The node state of C_8 is derived through reasoning based on the node states of $C_1 \sim C_2$.

These two steps belong to different reasoning types and should therefore be designed with appropriate reasoning algorithms separately. Based on the characteristics of target threat and attack advantage, the first-step decision reasoning algorithm for the FCM model is designed as follows:

$$\begin{cases} c_i(t+1) = f\left(\sum_{\substack{i=1 \\ j\neq i}}^{n} c_j(t) w_{ji}\right) \\ f(x) = \dfrac{x-b}{a-b} \end{cases} \tag{3-7}$$

Here, $a = \sum w_{ij}^{+}$; $b = \sum w_{ij}^{-}$, w_{ij}^{+} represent positively influenced weights, and w_{ij}^{-} represents negatively influenced weights.

This reasoning algorithm exhibits the following characteristics:

(1) The reasoning results fall within the range of [0,1], which are positively correlated with the comprehensive influence on the node and can be dynamically adjusted.

(2) When the input sequence is $C_3 \sim C_7 = \begin{bmatrix} 1 & 1 & 1 & 1 & 0 \end{bmatrix}$ (i.e., all fuzzy nodes take the maximum fuzzy values unfavorable to attack), the attack advantage C_1 is minimized, while the target threat level C_2 is maximized, which is consistent with subjective experience.

(3) When the input sequence is $C_3 \sim C_7 = \begin{bmatrix} 0 & 0 & 0 & 0 & 1 \end{bmatrix}$ (i.e., the fuzzy nodes select the minimum fuzzy values affecting the target threat level), the attack advantage C_1 is maximized, while the target threat level C_2 is minimized, which is consistent with subjective experience.

Hence, the reasoning algorithm given earlier is applicable to the decision reasoning of nodes $C_1 \sim C_2$.

$$\begin{cases} C_1 = \dfrac{\sum_{i=1}^{8}(C_i \times w_{i1}) + \sum_{i=3}^{6}|w_{i1}|}{\sum_{i=1}^{8}|w_{i1}|} \\ C_2 = \dfrac{\sum_{i=1}^{8}(C_i \times w_{i2})}{\sum_{i=1}^{8}|w_{i2}|} \end{cases} \tag{3–8}$$

Based on the relational characteristics between auxiliary decision-making nodes C_8 and $C_1 \sim C_2$, the second-step decision reasoning algorithm is designed as follows [62]:

$$c_i^{t+1} = \begin{cases} c_i^t + s_i^t(1 - c_i^t), & c_i^t \geq 0, s_i^t \geq 0 \\ c_i^t + s_i^t(1 + c_i^t), & c_i^t < 0, s_i^t < 0 \\ c_i^t + \dfrac{1}{2}s_i^t, & \text{otherwise} \end{cases} \tag{3–9}$$

Here, $s_i^t = \sum c_j w_{ji}$. To validate the rationality of Equation (3–9), an auxiliary decision-making FCM model for nodes $C_1 \sim C_2 \sim C_8$ is established, as shown in Figure 3.8.

An increase in C_1 will cause a decrease in C_8, while an increase in C_2 will cause an increase in C_8 and simultaneously cause a decrease in C_1. An increase in C_8 will cause an increase in C_1 while causing a decrease in

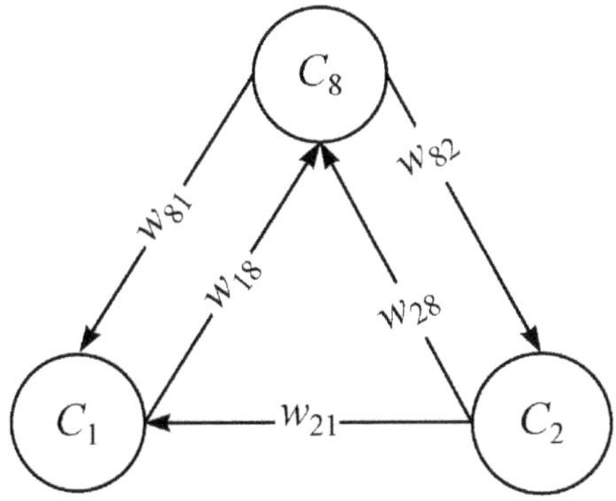

FIGURE 3.8 Auxiliary decision-making FCM model

C_2. Based on the aforementioned rules, the adjacency weight matrix is designed as follows:

$$W = \begin{bmatrix} 0 & 0 & -0.1 \\ -0.3 & 0 & 0.7 \\ 0.2 & -0.7 & 0 \end{bmatrix} \tag{3–10}$$

The simulation results of the iterative reasoning are shown in Figure 3.9.

Figure 3.9(a) shows the node state evolution diagram of the auxiliary decision-making FCM model, illustrating the evolution of key nodes. The state difference between nodes C_1 and C_2 represents our relative advantage value, as shown in Figure 3.9(b). The minimum jamming intensity is 0.4 units. Thus, the state of the auxiliary decision-making node C_8 (required jamming intensity) dynamically changes according to the relationship between node C_1 (our attack advantage) and node C_2 (target threat level). When the relative advantage decreases, the output of C_8 increases, resulting in a relative reduction of the target threat level. When the relative advantage is strong, the output of C_8 correspondingly decreases, and the target threat level relatively increases. Therefore, the auxiliary decision-making model in Figure 3.8 effectively simulates the suppressive effect of jamming on the target threat; meanwhile, the jamming intensity can be dynamically adjusted based on the target threat, thereby validating the rationality of using Equation (3–9) as the reasoning algorithm.

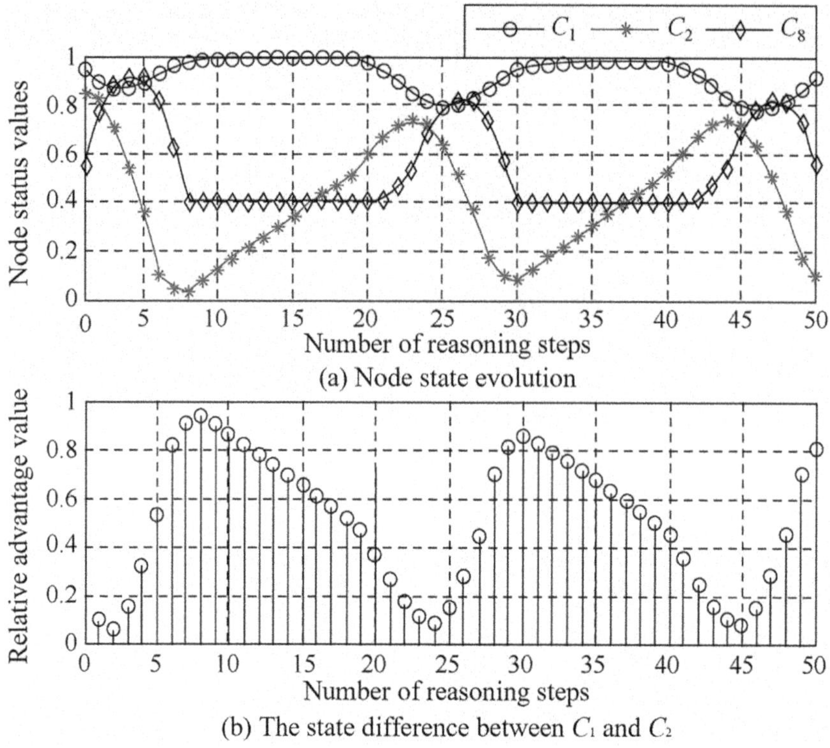

FIGURE 3.9 The simulation results of the iterative reasoning

3.4.2.3 Environmental Variable Definition and Model Parameter Adjustment

Three environmental variables affect the UAV's autonomous decision-making process:

(1) V_e: Weather, corresponding to a fuzzy value range of [0,1]. Generally, weather conditions affect information acquisition equipment to some extent. Worse weather conditions correspond to a smaller fuzzy value, resulting in poorer information acquisition capabilities for both enemy and friendly forces. The negative impact is greater on the defensive side, thereby increasing the advantage of the attacking side.

(2) V_1: Jamming intensity of our side, corresponding to a fuzzy value range of [0,1]. Higher jamming intensity corresponds to a larger fuzzy value, thereby increasing our attack advantage.

(3) V_f: Reception of support information from friendly forces, corresponding to a fuzzy value range of [0,1]. This primarily includes

support information from different platforms such as other formations of our side and ground control stations. A larger fuzzy value indicates a more significant advantage for our side.

After introducing V_e, V_l, and V_f, the specific adjustment relationships are derived based on the dynamic adjustment algorithm for the adjacency weight matrix W in Equation (3–5) as follows [61]:

$$w_{ij}^t = f_\omega^x\left(w_{ij}^0, V_k, a_k\right) = \begin{cases} w_{ij}^0(1 + a_k V_k), & -1 \le w_{ij}^t \le 1 \\ -1, & w_{ij}^t < -1 \\ 1, & w_{ij}^t > 1 \end{cases} \tag{3–11}$$

Here, w_{ij}^0 are the adjacency weights requiring adjustment under the influence of environmental variables, and w_{ij}^t represents the adjusted weights.

Thus, a DFCM model for UAV autonomous attack decision-making, considering the dynamic influence of environmental factors, can be established, as shown in Figure 3.10, along with the adjustment relationships for dynamic adjacency weights, as provided in Table 3.10.

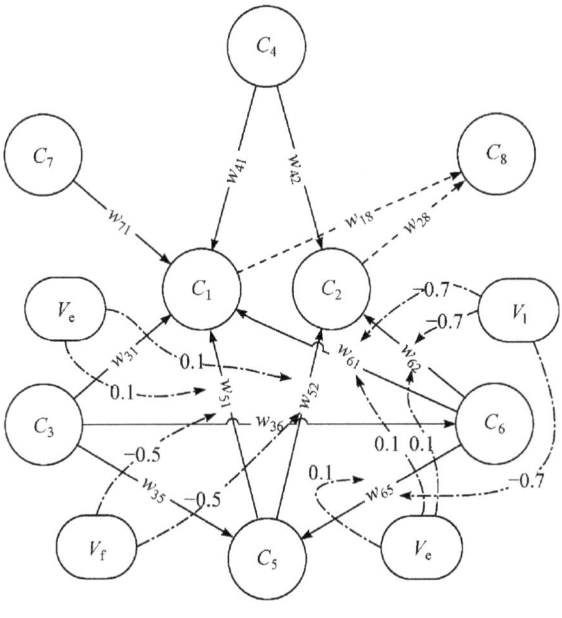

FIGURE 3.10 DFCM model for autonomous attack decision-making of UAV

TABLE 3.10 Adjustment Relationship Of Dynamic Adjacency Weights

w_{ij}	Adjustment of w_{ij} Based on V_k	
w_{51}	$f_{\omega}^{x}(w_{51},V_e,0.1)$	$f_{\omega}^{x}(w_{51},V_f,-0.5)$
w_{52}	$f_{\omega}^{x}(w_{52},V_e,0.1)$	$f_{\omega}^{x}(w_{52},V_f,-0.5)$
w_{61}	$f_{\omega}^{x}(w_{61},V_e,0.1)$	$f_{\omega}^{x}(w_{61},V_1,-0.7)$
w_{62}	$f_{\omega}^{x}(w_{62},V_e,0.1)$	$f_{\omega}^{x}(w_{62},V_1,-0.7)$
w_{65}	$f_{\omega}^{x}(w_{65},V_e,0.1)$	$f_{\omega}^{x}(w_{65},V_1,-0.7)$

3.4.3 Limited-Intervention Decision Model for Collaborative Attack in Manned Aircraft

3.4.3.1 Attack Decision Output Model for UAV

During the autonomous attack decision-making by the UAV, the battlefield environment is transformed into our attack advantage (C_1) and the target threat level (C_2). The decision outcome is dynamically adjusted based on the states of C_1 and C_2 [63]. An increase in our attack advantage facilitates attack, while an enhancement of the target threat hinders attack, as shown in Figure 3.11. When the output of the UAV's autonomous attack decision-making is uncertain, intervention from the manned aircraft is triggered.

$w(c_1)$ and $w(c_2)$ represent the dynamic weights of the states of nodes C_1 and C_2, respectively. The dynamic adjacency weights are designed as follows:

$$w(c_1)=\begin{cases}1, & c_1>0.8\\ c_1, & 0.4\le c_1\le 0.8\\ -1, & c_1<0.4\end{cases} \tag{3–12}$$

$$w(c_2)=\begin{cases}-1, & c_2>0.7\\ -c_2, & 0.35\le c_2\le 0.7\\ 1, & c_2<0.35\end{cases} \tag{3–13}$$

The reasoning decision function for the attack decision node is designed as follows:

$$\text{Decision Function}=S\left[w(c_1)+w(c_2)\right] \tag{3–14}$$

$$S(x)=\begin{cases}-1, & x\le\delta\\ 0, & \delta<x<\varepsilon\\ 1, & x\ge\varepsilon\end{cases} \tag{3–15}$$

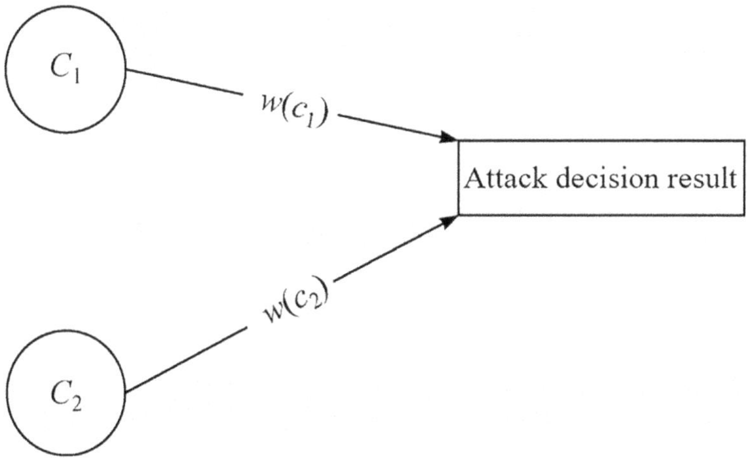

FIGURE 3.11 FCM model for attack decision-making

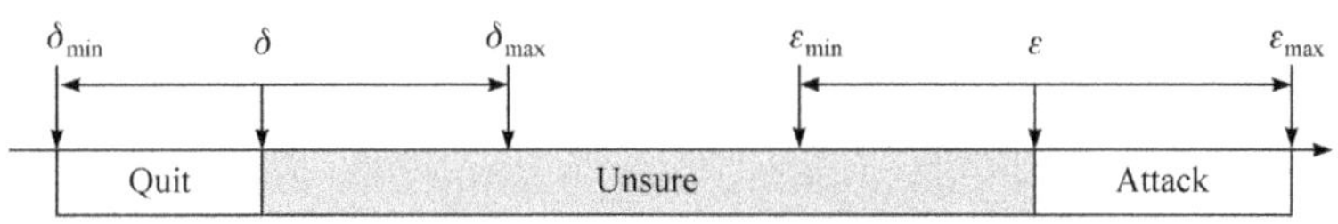

FIGURE 3.12 Illustration of the intuitive meaning of decision threshold parameters

$$\text{Decision Result} = \begin{cases} \text{Quit}, & S(x) = -1 \\ \text{Unsure}, & S(x) = 0 \\ \text{Attack} & S(x) = 1 \end{cases} \tag{3–16}$$

Here, δ and ε are the threshold parameters for the decision outcome, with their intuitive meanings illustrated in Figure 3.12. The values of δ and ε determine the decision output. These values can be predefined based on expert experience or learned from data samples and can also be dynamically adjusted based on the changes in the battlefield environment. *Decision Result* represents the autonomous decision outcome of the UAV, where *Quit* indicates the decision result "withdraw", *Unsure* indicates "uncertain", and *Attack* indicates "attack". δ_{max} and δ_{min}, ε_{max} and ε_{min} are the upper and lower limits of δ and ε, respectively.

The confidence level of the decision reasoning outcome is calculated as follows:

$$\begin{cases} A_{\text{confidence}} = \left| \dfrac{r - \varepsilon_{\min}}{\varepsilon_{\max} - \varepsilon_{\min}} \right| \\ Q_{\text{confidence}} = \left| \dfrac{r - \delta_{\max}}{\delta_{\max} - \delta_{\min}} \right| \end{cases} \tag{3–17}$$

Here, $A_{\text{confidence}}$ represents the confidence level when the decision result is Attack. $Q_{\text{confidence}}$ represents the confidence level when the decision result is Quit $r = w(c_1) + w(c_2)$. The closer the reasoning value for the Attack decision result is to $\varepsilon_{\max}$, the higher the confidence level of the decision result. The closer the reasoning value for the Quit decision result is to $\delta_{\min}$, the higher the confidence level of the decision result.

3.4.3.2 Limited-Intervention Decision Model for Collaborative Attack in Manned Aircraft

Under the consent management mode, if the UAV encounters difficulties in autonomous attack decision-making, it adjusts its autonomy level and requests intervention from the manned aircraft. During the UAV's autonomous attack decision-making process, the manned aircraft's intervention methods are the most diverse, and the scenarios are the most complex. The limited-intervention decision-making model for manned aircraft collaborative attack is shown in Figure 3.13.

3.4.3.2.1 Direct Control Considering the mission requirements for ground attack, when the mental workload of the manned aircraft operator is low, priority is given to conducting command operations on the UAV, at which point the UAV reduces its autonomy level. When the UAV is in the operator decision mode, it provides mission planning parameters (i.e., our attack advantage and target threat), while the manned aircraft makes attack decisions and authorizes the UAV to deploy weapons. When the UAV is in the instruction control mode, it provides battlefield situational information, and the manned aircraft operator analyzes the attack advantage and target threat to make attack decisions.

3.4.3.2.2 Intervention Environment When the UAV encounters decision-making difficulties, the auxiliary decision-making node simultaneously

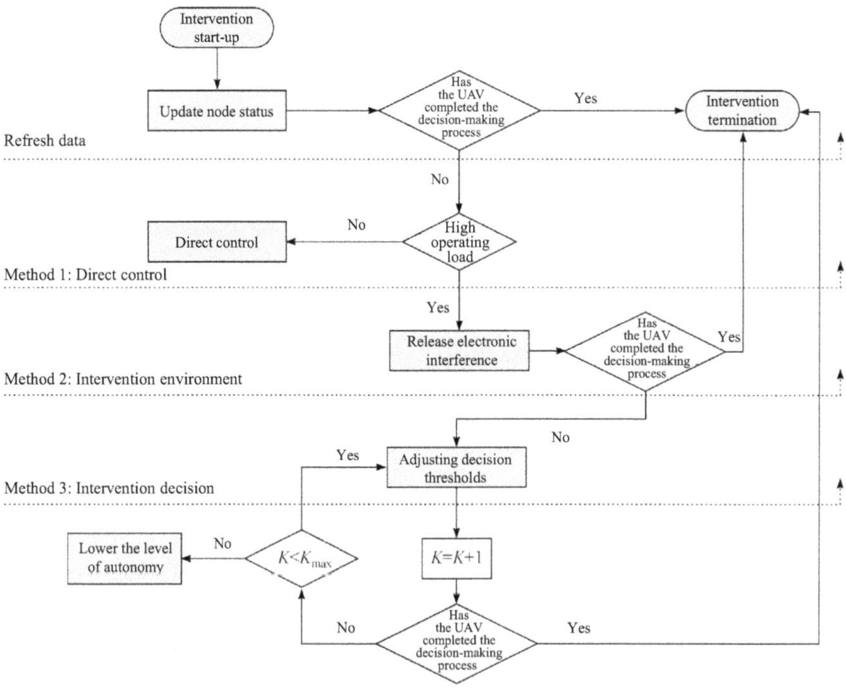

FIGURE 3.13 Limited-intervention decision model for collaborative attack in manned aircraft

sends an intervention request and provides auxiliary information for the manned aircraft's intervention. According to the definition of the auxiliary decision-making node C_g, when the UAV requests intervention from the manned aircraft, it synchronously transmits information suggesting the intensity of jamming to be deployed. The manned aircraft's deployment of jamming can alter the battlefield situation, utilizing the auxiliary decision-making FCM model to control the states of nodes C_1 and C_2 while simultaneously updating the UAV's data and information regarding the target. This thereby influences the output of the UAV's autonomous decision-making results, achieving limited-intervention decision-making.

3.4.3.2.3 Intervention Decision-Making When difficulties arise in the autonomous decision-making reasoning of the UAV, the manned aircraft can intervene through decision-making intervention methods such as supplementing target priority information and adjusting decision threshold parameters to collaboratively complete attack decisions with the UAV.

The principles for adjusting the threshold parameters of the UAV's autonomous decision-making reasoning are as follows:

(1) If the target threat is high, increase parameter δ to make the UAV tend to choose Quit.

(2) If the target threat is low, decrease parameter ε to make the UAV tend to choose Attack.

(3) If the target task priority is high, decrease parameter ε to make the UAV tend to choose Attack.

(4) If the target task priority is low, increase parameter δ to make the UAV tend to choose Quit.

Based on these four principles, the following rules are derived:

If the task priority is low, then Quit.

If the task priority is high and the threat is high, then Unsure.

If the task priority is high and the threat is low, then Attack.

According to the adjustment rules, when the task priority is low, the attack decision result of the UAV is set to quit. When the task priority is high, the adjustment functions for δ and ε are as follows:

$$\begin{cases} \delta(T)=\begin{cases}\delta_0, & T \le 2T_0 \\ e^{-T+2T_0}+\delta_0-1, & T>2T_0\end{cases} \\ \varepsilon(T)=AT^2+BT+C \\ A=\dfrac{\varepsilon_0(1-\varepsilon_{\min})}{T_0^2} \\ B=\dfrac{2\varepsilon_0(\varepsilon_{\min}-1)}{T_0} \\ C=\varepsilon_0 \end{cases} \tag{3–18}$$

Here, δ_0 and ε_0 are the decision threshold parameters for autonomous decision-making. T is the threat of the target. T_0 is the minimum threat value. The adjustment range for δ is $[\delta_0, \delta_{\min}]$ while for ε is $[\varepsilon_{\min}, \varepsilon_{\max}]$.

Assuming a set of parameters, a simulation of the adjustment functions for δ and ε as the target threat varies is conducted, with the results shown in Figure 3.14.

As the target threat level increases, the variation curves of parameters δ and ε divide the decision result plane into three regions: Attack, Unsure, and Quit. The autonomous decision-making results of the UAV FCM model fall into different regions, corresponding to distinct decision outcomes. After activating the decision-threshold intervention method, high target priority coupled with relatively low threat leads the UAV's autonomous decision-making to favor an Attack action. As the target threat level gradually increases, the attack risk of the UAV rises, and its autonomous decision-making tends to favor Unsure. At this point, the UAV reduces its autonomy level to await further intervention from the manned aircraft.

Given the characteristics of UAV ground attack decision-making, the final decision outcome can only be to attack the target or abort the attack. Hovering, as a transitional state, occurs only when decision-making difficulties arise. The initial autonomy level for the UAV during ground attack is set to the consent management mode. When the decision result of the UAV is uncertain, it cannot form mission planning results. At this point,

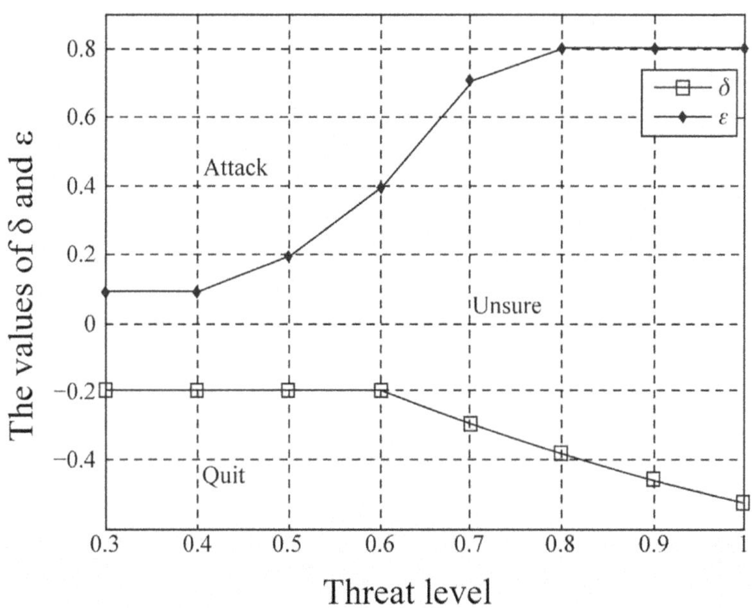

FIGURE 3.14 Decision threshold adjustment

the UAV reduces its autonomy level, transmits mission parameters to the manned aircraft, and requests intervention. If no intervention feedback is received from the manned aircraft within the waiting period, it chooses to abort the attack.

3.5 SIMULATION AND ANALYSIS OF THE LIMITED-INTERVENTION COLLABORATIVE DECISION-MAKING BETWEEN MANNED AIRCRAFT AND UAV

3.5.1 Simulation Process and Selection of Analysis Metrics

This section focuses on simulations of three key phases during the mission process of the UAV: obstacle avoidance, threat avoidance, and attack decision-making. Corresponding evaluation metrics are selected for analysis, with the simulation and analysis process illustrated in Figure 3.15.

Based on the characteristics and applicability of UAV autonomous decision-making, the selected evaluation metrics are as follows:

3.5.1.1 Mission Complexity

A mission is composed of a sequence of actions, and the mission complexity is related to the complexity of the actions. The action complexity function is defined as follows:

$$d(a) = \begin{cases} d_1 = 1, & a \in A_{\mathrm{H}} \\ d_2 = \dfrac{1}{2}, & a \in A_{\mathrm{I}} \\ d_3 = 0, & a \in A_{\mathrm{L}} \end{cases} \tag{3–19}$$

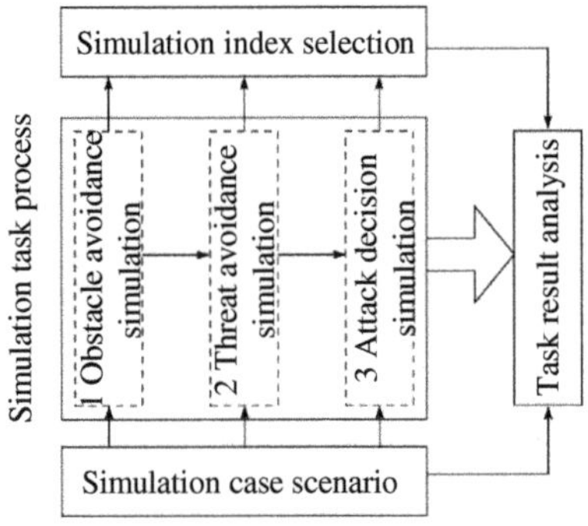

FIGURE 3.15 Simulation and analysis process

Here, $d(a)$ is the complexity of action a. When a mission consists of multiple action sequences, its complexity can be expressed as follows:

$$D(t) = 1 - \frac{1}{d_1 N_H + d_2 N_I + d_3 N_L} \tag{3-20}$$

$D(t)$ denotes the complexity of task t. N_H represents the number of high-level actions contained in the task action sequence. N_I represents the number of mid-level actions contained in the task action sequence. N_L represents the number of low-level actions contained in the task action sequence. The complexity of task t can be further expressed as follows:

$$D(t) = 1 - \frac{1}{N_H + 0.5 N_I} \tag{3-21}$$

The complexity $D(t) \in [0,1]$ of task t satisfies $D(t) = 0$, indicating that task t is extremely simple, while $D(t) = 1$ indicates that task t is highly complex. The more complex the tasks handled by the UAV, the higher its autonomy level, thereby simplifying the operator's tasks. If $D(t)$ is the complexity of tasks executed by the UAV, the complexity of tasks for the operator is as follows:

$$\hat{D}(t) = 1 - D(t), \qquad \hat{D}(t) \in [0,1] \tag{3-22}$$

3.5.1.2 Number of Interactions

The number of interactions reflects the information exchange between the manned aircraft and the UAV. Interactions between them always involve intervention by the manned aircraft and adjustments to the autonomy level of the UAV. The number of interactions between the manned aircraft and the UAV affects the operator's SA level. Excessive interactions can sharply increase operator pressure, leading to an "overload" phenomenon, while insufficient interactions can result in the "human-out-of-the-loop" phenomenon [64–66]. The counting of interactions can be completed during the mission simulation process.

3.5.1.3 Operator SA Level

The operator's SA level, along with the number of interactions between the manned aircraft and the UAV, is related to the autonomy level of the UAV.

Generally, under the consent management mode, the operator's SA of the battlefield is the highest, followed by manual operation. While under the exception management mode, the operator's SA level is the lowest. From the operator's perspective, there is a preference for the UAV to operate in the consent management mode as much as possible. Hence, the SA capability of the UAV at various autonomy levels can be qualitatively described as follows: Consent management mode > Instruction operation > Operator decision mode > Exception management mode. Therefore, the evaluation formula for the operator's SA level, based on the UAV's autonomy level and the number of human-machine interactions, can be expressed as follows:

$$\mathrm{SA} = \sum_{i=1}^{4} \frac{\mathrm{Ca}_i}{\mathrm{Cn}} \cdot \mathrm{wg}_i \tag{3-23}$$

Here, SA denotes the operator's situation awareness level. Ca_i is the number of occurrences of autonomy level *i*. Cn represents the total number of interactions during the mission. wg_i represents the weight of the influence of autonomy level *i* on SA.

3.5.1.4 Task Completion Degree

The task completion degree, to a certain extent, reflects the effectiveness of the UAV in accomplishing missions and serves as the most direct metric for evaluating the execution efficacy of the limited intervention-based collaborative decision-making mechanism between the manned aircraft and the UAV. The calculation method for the task completion degree is defined as follows:

$$\text{Task Completion Degree} = \frac{n}{N} \times 100\% \tag{3-24}$$

where *n* is the number of completed subtasks, and *N* is the total number of subtasks.

3.5.2 Simulation and Analysis of Collaborative Obstacle Avoidance Decision-Making for Manned Aircraft and UAV

Three simulation scenarios for the UAV obstacle avoidance task are categorized into two primary types: normal communication and communication interruption. Under the normal communication conditions, the scenarios are further divided into known obstacles and unknown obstacles, as detailed in Table 3.11.

TABLE 3.11 Three Simulation Scenarios for UAV Obstacle Avoidance Tasks

Assumptions	Obstacle Types	Obstacle Distance	Communication Status	Initial Autonomy Level
1	Known obstacles	Far	Normal	Consent management model
2	Unknown obstacles	Close	Normal	Consent management model
3	Unknown obstacles	Far	Interruption	Consent management model

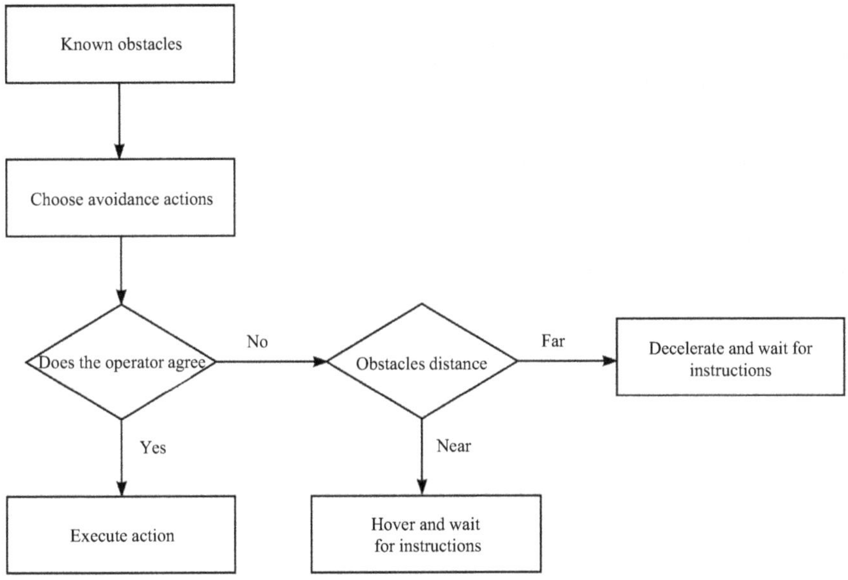

FIGURE 3.16 Avoidance process of UAV for known obstacles

For Scenario 1, the obstacle type is known, and the UAV possesses the necessary knowledge for obstacle avoidance. Therefore, no intervention from the manned aircraft is required, and the UAV autonomously completes the obstacle avoidance decision-making process. Under the approval management mode, operator consent must be obtained before execution, as shown in Figure 3.16.

For Scenario 2, the obstacle type is unknown, and the UAV lacks the knowledge required for obstacle avoidance. It sends obstacle image information to the manned aircraft and requests intervention. After receiving the obstacle information transmitted by the UAV, the manned aircraft supplements the avoidance type of the obstacle through the intervention interface, assisting the UAV in completing obstacle avoidance. Throughout the avoidance process, the UAV remains under the approval management mode. Within this level, the autonomy level of the UAV is adjusted horizontally by reducing the complexity of the selected actions, as shown in Figure 3.17.

For Scenario 3, when the UAV encounters an unknown obstacle and requests intervention from the manned aircraft, communication is interrupted, resulting in the inability to receive intervention information from the manned aircraft. In this case, the autonomy level adjustment of the UAV includes two dimensions: (1) During the request for intervention, a horizontal adjustment of the autonomy dimension is performed to narrow the action range and reduce the complexity of actions. (2) If no intervention feedback is received from the manned aircraft within a certain time limit, a vertical adjustment of the autonomy level is conducted to lower the autonomy level and await the operator's decision results and operational commands. If no feedback is received from the manned aircraft beyond the time limit, the UAV chooses self-preservation and returns to the base, as shown in Figure 3.18.

Through simulation, the statistics of various metrics in the UAV's obstacle-avoidance decision-making process can be obtained, as shown in Table 3.12.

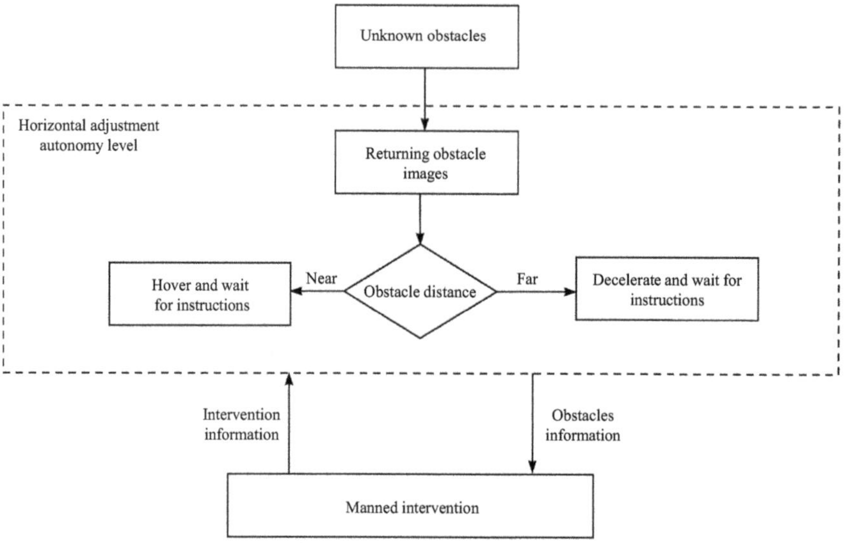

FIGURE 3.17 Avoidance process of UAV for unknown obstacles

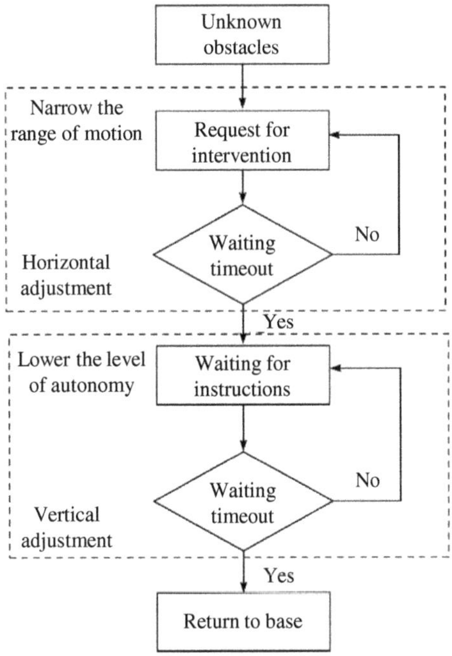

FIGURE 3.18 Obstacle avoidance process of UAV under communication interruption

TABLE 3.12 Statistics of Metrics for the UAV Obstacle-Avoidance Task Process

Assumptions	No. of Interactions	Execution Sequence of Actions (Autonomy Level)	Task Status
1	1	Obstacle avoidance (3)	Finish
2	2	Hovering and waiting for instructions (3) → Obstacle avoidance (3)	Finish
3	0	Deceleration, wait instruction (2) → Return (2)	Unfinished

In Table 3.12, the autonomy level at which each action is executed is indicated in parentheses following the action name. Analysis of the simulation results reveals that, for known obstacles, the UAV can fully utilize its autonomous capabilities to complete obstacle avoidance independently. For unknown obstacles, the UAV requires intervention from the manned aircraft to accomplish avoidance, leveraging both the autonomous execution capabilities of the UAV and the analytical judgment of the manned aircraft operator. For extreme situations like communication interruption, the UAV can adjust its autonomy level to ensure its own safety while awaiting communication recovery, opting for self-preservation when necessary. Thus, it can be observed that the limited-intervention decision-making mechanism effectively integrates the strengths of both manned aircraft and UAV, demonstrating strong adaptability and responsiveness to different obstacle types and environmental conditions.

3.5.3 Simulation of a Case Study on UAV Threat Avoidance Mission

Three simulation scenarios for the UAV threat avoidance mission are categorized into two primary types: normal communication and communication interruption. Under normal communication conditions, the scenarios are divided into known and unknown threats, as specified in Table 3.13.

For Scenario 1, the UAV possesses avoidance knowledge for known threats and can autonomously complete the threat avoidance decision-making process. For high-value threat sources, the UAV may either choose to avoid them or request intervention from the manned aircraft to destroy the threat source through fire strikes, as shown in Figure 3.19.

For Scenario 2, the UAV cannot autonomously complete the threat avoidance decision-making process and must request intervention from the manned aircraft. Simultaneously, it transmits relevant information about the threat source back to the manned aircraft. Based on this, the manned aircraft operator can conduct a comprehensive analysis and judgment and assist the UAV in completing the threat avoidance mission through the intervention interface. The process of the UAV avoiding unknown threats is shown in Figure 3.20.

The UAV transmits information regarding the unknown threat to the manned aircraft and requests intervention, while simultaneously awaiting

TABLE 3.13 Three Simulation Scenarios for UAV Threat Avoidance Missions

Assumptions	Threat Types	Threat Value	Threat Distance	Communication Status	Initial Autonomy Level
1	Known threats	High value	Medium	Normal	Consent management model
2	Unknown threat	Unknown	Relatively far	Normal	Consent management model
3	Unknown threat	Unknown	Relatively far	Interruption	Consent management model

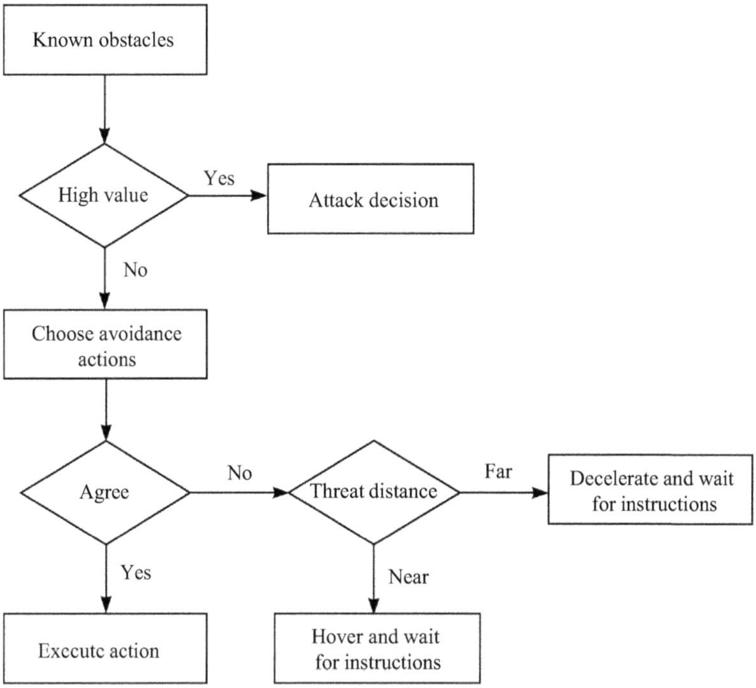

FIGURE 3.19 Avoidance process for known threats

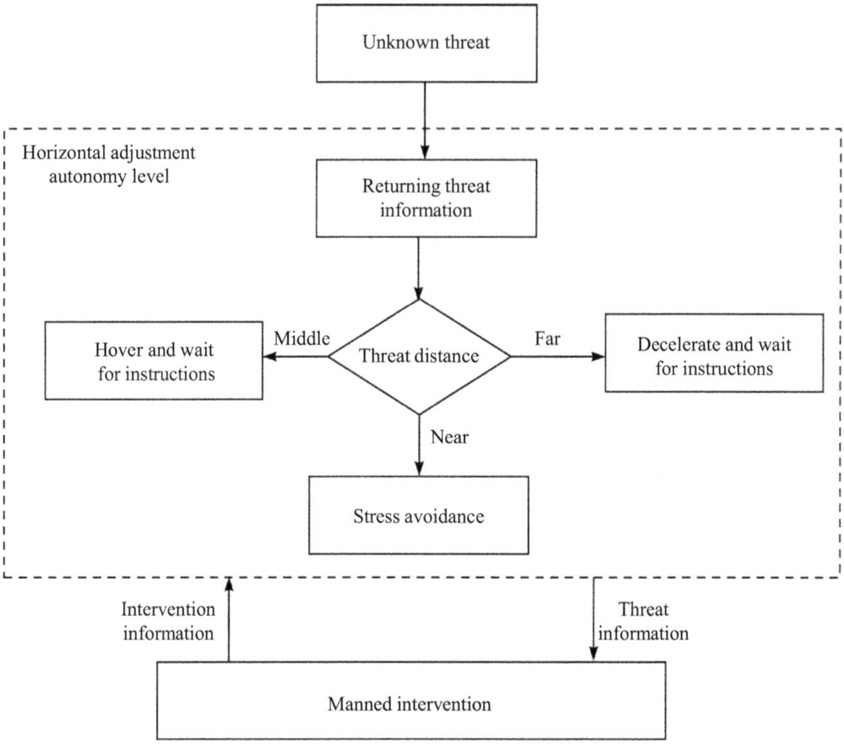

FIGURE 3.20 Avoidance process of UAV for unknown threats

intervention instructions from the manned aircraft. The manned aircraft operator conducts a comprehensive analysis of the threat information provided by the UAV and selects an intervention method. The primary intervention methods available to the manned aircraft here are threefold: (1) directly controlling the UAV via commands to complete avoidance, (2) supplementing threat source information to assist the UAV in completing avoidance, (3) for highly valuable threat sources, the manned aircraft directly attacks the threat or controls the UAV to attack the threat. In Scenario 2, the UAV remains under the approval management mode throughout and, within this level, performs a horizontal adjustment of its autonomy level by reducing the complexity of the selected actions. The manned aircraft assists the UAV in completing threat avoidance by supplementing threat source information.

For Scenario 3, the UAV cannot autonomously complete the threat avoidance decision-making process while simultaneously being unable to obtain intervention information from the manned aircraft. In this case, the UAV chooses to adjust its autonomy level. If no intervention information is received from the manned aircraft beyond the specified time limit, it opts for self-preservation and returns to the base. The threat avoidance process of the UAV under communication interruption is shown in Figure 3.21. Like obstacle avoidance, the threat avoidance process of the UAV under communication interruption involves adjustments to the autonomy level in two dimensions.

Through simulation, various metrics in the UAV's threat avoidance decision-making process can be statistically analyzed, as shown in Table 3.14.

Analysis of the simulation results indicates that for known threats, the UAV can fully utilize its autonomous capabilities to independently complete threat avoidance or eliminate threats by attacking threat sources. For unknown threats, the UAV must complete avoidance under the guidance of the manned aircraft operator, leveraging both the autonomous execution capabilities of the UAV and the analytical judgment of the manned aircraft operator. For extreme situations such as communication interruption, the UAV can adjust its autonomy level to ensure its own safety while awaiting communication recovery. This demonstrates that the limited-intervention decision-making mechanism effectively integrates the characteristics of both unmanned and manned aircraft, exhibiting strong adaptability to different threat types and environmental conditions.

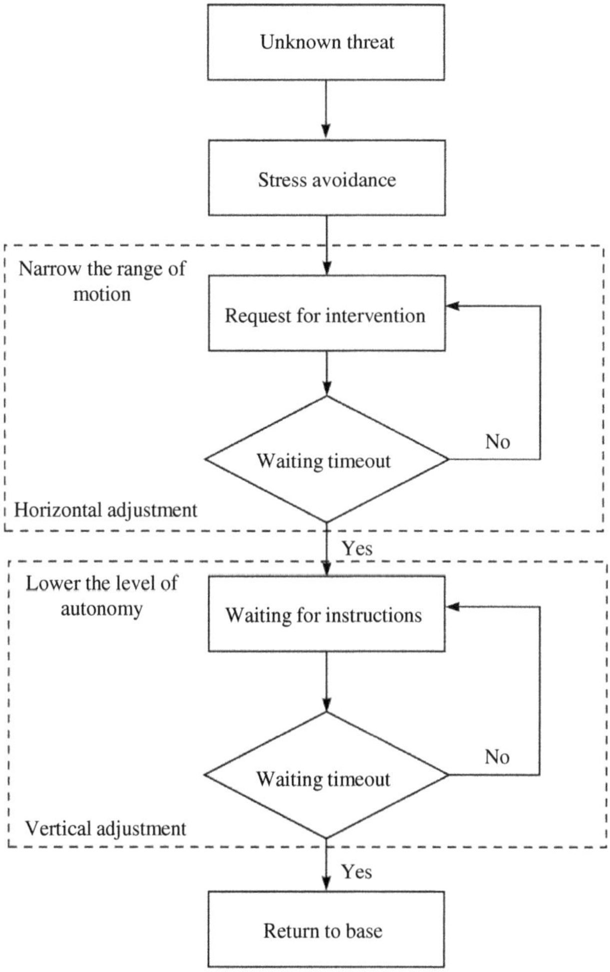

FIGURE 3.21 Threat avoidance process of UAV under communication interruption

TABLE 3.14 Statistics of Process Metrics for UAV Threat Avoidance Missions

Assumptions	No. of Interactions	Execution of the Action Sequence (autonomy level)	Task Status
1	2	Attack decision (3)/Obstacle avoidance (3)	Finish
2	2	Deceleration wait instruction (3) → Obstacle avoidance (3)	Finish
3	0	Reactive avoidance (3) → Hover and waiting for instructions (2) → Return (2)	Unfinished

TABLE 3.15 UAV Attack Decision Target Scenario

Target	C_3	C_4	C_5	C_6	C_7
1	0.1	0.1	0.1	0.1	45
2	0.5	0.5	0.5	0.5	65
3	0.4	0.5	0.5	0.6	50
4	0.8	0.6	0.6	0.8	70
Data types	Fuzzy value	Fuzzy value	Fuzzy value	Fuzzy value	True value (km)

3.5.4 Simulation of a Case Study on UAV Target Attack Mission

The target attack process of UAV is relatively complex, with intervention methods during attack decision-making being relatively diverse. Autonomous attack decision-making by UAV involves the fuzzification of battlefield situations. Here, the fuzzification results of nodes in the FCM model representing the situation will be directly adopted as scenarios, as shown in Table 3.15.

Table 3.15 selects four representative target groups for attack decision-making simulation. Analysis indicates that Target 1 exhibits low terrain complexity, low attack difficulty, poor enemy air defense firepower, poor radar status, and maximum distance advantage, making it the most favorable for attack. Target 4 features complex terrain, high attack difficulty, strong enemy air defense firepower, relatively good radar status, and weak distance advantage, making it unfavorable for attack. The node states of Target 2 and Target 3 fall between those of Target 1 and Target 4. Due to the specific nature of UAV ground attack missions, the final decision outcome can only be "Attack" or "Quit". The hovering-waiting state is merely a transitional state when the autonomous decision result is "Unsure". When the UAV's autonomous decision result is "Unsure", intervention from manned aircraft is required to complete collaborative attack decision-making.

3.5.4.1 Interface of the UAV Attack Decision Supervision and Control Simulation System

As shown in Figure 3.22, the simulation interface for UAV attack decision-making with manned aircraft limited-intervention supervision and control is divided into four main modules:

(1) Input Module: Includes terrain complexity, target attack difficulty, enemy air defense intensity, enemy radar status, target distance, target priority, and environmental impact factors (V_e, V_l, V_f).

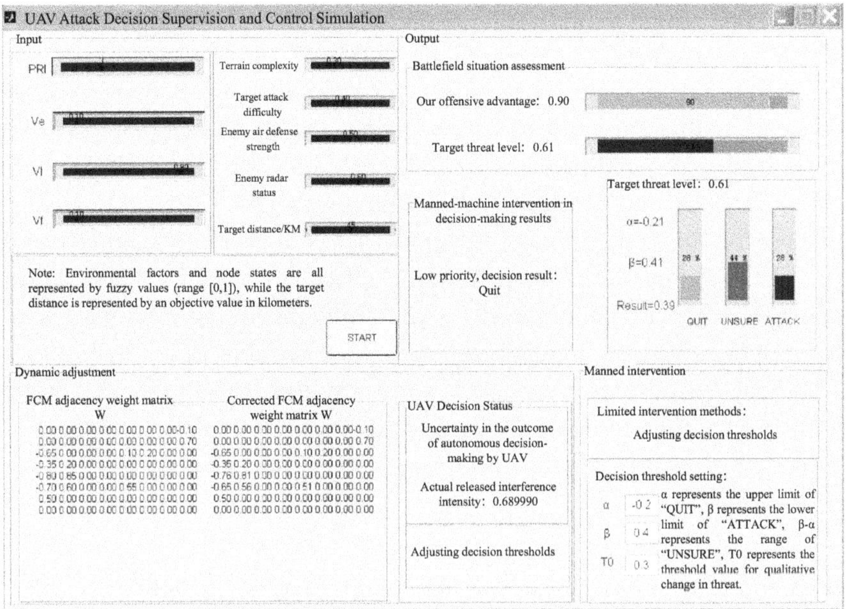

FIGURE 3.22 Simulation interface for UAV attack decision-making and manned aircraft limited-intervention supervision and control

(2) Output Module: Comprises battlefield situation assessment (i.e., our attack advantage and target threat level), the final decision result of manned aircraft and UAV, and the adjustment results of decision thresholds by the manned aircraft.

(3) UAV Decision Status Module: Displays the autonomous decision results of the UAV and the method of intervention requested by the manned aircraft.

(4) Manned Aircraft Intervention Module: Shows the intervention methods and decision threshold settings of the manned aircraft.

After the attack decision-making task commences, the UAV performs autonomous decision-making based on the battlefield situation. When difficulties arise in autonomous decision-making, the UAV transmits the battlefield situation information to the manned aircraft. Based on the received battlefield situation information, the manned aircraft intervenes in the UAV, thereby completing the collaborative attack decision-making process between the manned aircraft and the UAV.

3.5.4.2 Simulation of Dynamic Adjustment of Adjacency Weight Matrix

The adjacency weight matrix of the UAV's FCM-based reasoning and decision-making model represents the causal relationships between concepts and the weights of causal influences, embodying expert prior knowledge and database support. The dynamic adjustment of the adjacency weight matrix based on changes in external environmental variables reflects the dynamic reasoning characteristics of the UAV's autonomous decision-making FCM model. Assuming the fuzzy values of the external environmental variable V are all 0.1, the simulation of the dynamic adjustment of the adjacency weight matrix is shown in Table 3.16.

When multiple factors exert influence simultaneously, the result with greater influence is selected and bolded in the table. According to Table 3.16, the modified adjacency weight matrix is as follows:

$$\begin{bmatrix} 0 & 0 & 0 & 0 & 0 & 0 & 0 & -0.1 \\ 0 & 0 & 0 & 0 & 0 & 0 & 0 & 0.7 \\ -0.65 & 0 & 0 & 0 & 0.1 & 0.2 & 0 & 0 \\ -0.35 & 0.2 & 0 & 0 & 0 & 0 & 0 & 0 \\ \mathbf{-0.760} & \mathbf{0.807} & 0 & 0 & 0 & 0 & 0 & 0 \\ \mathbf{-0.651} & \mathbf{0.558} & 0 & 0 & \mathbf{0.511} & 0 & 0 & 0 \\ 0.5 & 0 & 0 & 0 & 0 & 0 & 0 & 0 \\ 0 & 0 & 0 & 0 & 0 & 0 & 0 & 0 \end{bmatrix}$$

TABLE 3.16 Simulation of Dynamic Adjustment of Adjacency Weight Matrix

w_{ij}	w^0_{ij}	The Impact of V_k	w^t_{ij}
w_{51}	−0.8	$f^x_\omega(w_{51}, V_{e1}, 0.1)$	−0.808
		$f^x_\omega(w_{51}, V_{f1}, -0.5)$	0.760
w_{52}	0.85	$f^x_\omega(w_{52}, V_{e1}, 0.1)$	0.858
		$f^x_\omega(w_{52}, V_{f1}, -0.5)$	0.807
w_{61}	−0.7	$f^x_\omega(w_{61}, V_{e1}, 0.1)$	−0.707
		$f^x_\omega(w_{61}, V_{l1}, -0.7)$	0.651
w_{62}	0.6	$f^x_\omega(w_{62}, V_{e1}, 0.1)$	0.606
		$f^x_\omega(w_{62}, V_{l1}, -0.7)$	0.558
w_{65}	0.55	$f^x_\omega(w_{65}, V_{e1}, 0.1)$	0.555
		$f^x_\omega(w_{65}, V_{l1}, -0.7)$	0.511

3.5.4.3 *Simulation of UAV Attack Decision-Making and Manned Aircraft Limited-Intervention Decision-Making*

The collaborative attack decision-making between manned aircraft and UAVs is conducted based on the simulation scenarios described in Table 3.15. Table 3.17 shows the autonomous decision-making simulation results of the UAV. The symbol "●" indicates the decision result "Attack", "◐" indicates "Unsure", "□" indicates "Quit", and "\" indicates that the corresponding cell is empty. The same symbols represent identical meanings in subsequent sections. It should be noted that when the interference intensity output by the manned aircraft is insufficient to intervene in the UAV's definite decision-making, the value of C_8 will be updated.

Table 3.18 shows the simulation results of the manned aircraft and UAV collaborative attack decision-making. The second column displays the

TABLE 3.17 Simulation Results of UAV Autonomous Decision-Making

Target	C_1	C_2	δ	ε	Reasoning Results	Decision Results	C_8	Confidence
1	0.92	0.10	−0.20	0.40	1.00	●	—	1.00
2	0.43	0.50	−0.20	0.40	−0.06	◐	0.92	—
3	0.57	0.54	−0.20	0.40	0.05	◐	0.69	—
4	0.25	0.67	−0.20	0.40	−1.00	□	—	0.82
Data type					Fuzzy values			

TABLE 3.18 Simulation Results of Manned Aircraft and UAV Collaborative Attack Decision-Making

Target	Autonomous Decision-Making of Unmanned Aerial Vehicles	Intervention Needs	Intervention Type	Additional Information	Collaborative Decision-Making
1	●	×	—	—	●
2	◐	√	¤	—	●
3	◐	√	¤ ↔	■	●
4	□	×	—	—	□

Note: "√" and "×" denote that intervention is required and not required, respectively. "■" and "□" denote high and low target/task priority, respectively. "¤" indicates the intervention mode of electronic jamming. "↔" indicates the intervention mode of adjusting the decision threshold.

autonomous decision-making results of the UAV from Table 3.17, while the last column presents the final collaborative decision-making results of the manned aircraft and UAV system.

For Target 2 and Target 3, the UAV's autonomous decision-making result is uncertain, triggering intervention. Table 3.19 shows the adjustment of autonomous decision-making parameters of the UAV under manned aircraft intervention. In Table 3.19, C_1^* and C_2^* represent our attack advantage and the inferred results of target threat after intervention, respectively. δ^* and ε^* are the adjusted decision thresholds after intervention, and the last column shows the inferred result value of the UAV after intervention.

As shown in Tables 3.17 and 3.18, Target 1 and Target 4 exhibit relatively distinct characteristics, allowing their engagement to be determined by the UAV's autonomous decision-making. For Target 2 and Target 3, however, the UAV's autonomous decision-making yields uncertain results, necessitating intervention from the manned aircraft for collaborative decision-making. This process is illustrated in Figure 3.23, where solid-line boxes represent decision-state transitions and dashed-line boxes denote specific intervention methods employed by the manned aircraft.

As shown in Tables 3.17 and 3.19, for Target 2, the advantage and target threat values are 0.43 and 0.50, respectively, resulting in an uncertain autonomous decision by the UAV. Upon intervention, based on recommendations from the UAV's auxiliary decision node, jamming intensity is applied. Following two jamming cycles by the manned aircraft, the recalculated advantage and target threat values shift to 0.86 and 0.57, respectively, enabling the UAV to reach a definitive autonomous decision outcome, which is "Attack".

For Target 3, our advantage and target threat values are 0.57 and 0.54, respectively, and the UAV's autonomous decision-making result is uncertain, initiating intervention. After applying jamming, our advantage and

TABLE 3.19 Adjustment of Autonomous Decision-Making Parameters of the UAV Under Intervention

Target	C_1^*	C_2^*	δ^*	ε^*	**Intervention Reasoning Outcome Value**
2	0.86	0.57	−0.20	0.40	0.43
3	0.79	0.55	−0.15	0.31	0.44
Data type	Fuzzy value	Fuzzy value	Fuzzy value	Fuzzy value	Fuzzy value

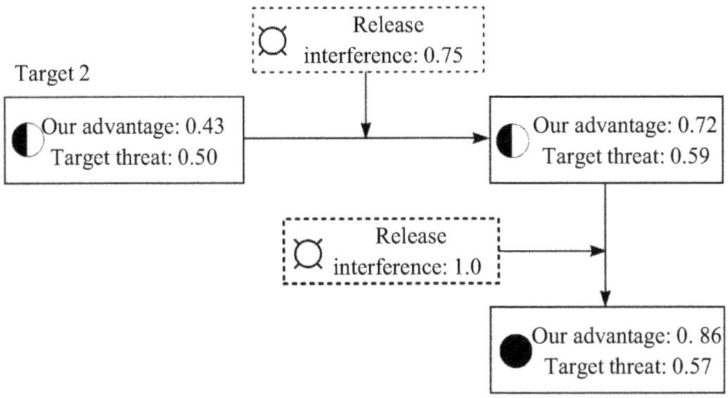

(a) Collaborative decision-making for objective 2

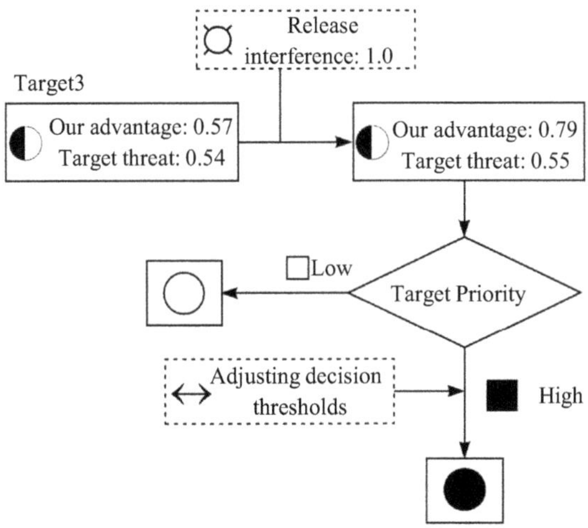

(b) Collaborative decision-making for objective 3

FIGURE 3.23 Collaborative decision-making process for Targets 2 and 3

target threat values become 0.79 and 0.55, respectively, yet the UAV still cannot provide a definite decision and requires supplementary target-priority information. For Target 3, if the target priority is low, the UAV's decision result is "Quit". In this scenario, the supplementary information specifies a high target priority. Therefore, the decision threshold is dynamically adjusted based on the target threat level. After adjustment, the UAV's decision simulation result becomes "Attack".

Through simulation, various metrics in the UAV target attack decision-making mission can be statistically analyzed, as shown in Table 3.20.

TABLE 3.20 Statistics of Process Metrics for UAV Target Attack Missions

Target Assumptions	Number of Interactions	Action Sequence (autonomy level)	Task Completion Status
1	1	Attack (3)	Finish
2	3	Hovering and waiting (3) → Attack (3)	Finish
3	3	Hovering and waiting (3) → Attack (3)	Finish
4	1	Return (3)	Unfinished

As can be seen from Table 3.20, for targets with relatively distinct characteristics such as Target 1 and Target 4, the UAV can autonomously complete attack decision-making. When the target threat and our attack advantage are close and the battlefield situation is ambiguous, the UAV experiences difficulties in autonomous decision-making. This necessitates leveraging the manned aircraft operator's cognitive analytical capabilities regarding the battlefield situation to complete the decision-making collaboratively through intervention in both environmental factors and the UAV's decision-making process.

3.5.5 Analysis and Conclusions of Simulation Results

To reflect the characteristics of the variable autonomy supervision and control mode, this subsection first summarizes the features of variable autonomy supervision and control during the subtask execution process. Subsequently, the subtasks are connected in series, and the task performance metrics of the UAV under command control mode, variable autonomy mode, and full autonomy mode are analyzed and compared, respectively, aiming to illustrate the advantages and disadvantages of each mode, as well as the applicable scenarios of the variable autonomy mode.

3.5.5.1 Analysis of Simulation Results for Subtasks

The simulation scenarios for obstacle avoidance, threat avoidance, and target attack tasks are designed from simple to complex. The analysis of the corresponding simulation results leads to the following conclusions:

(1) For simple tasks, the UAV can fully utilize its autonomy under the autonomy level adjustment mechanism to complete tasks independently, require fewer interactions, and minimize the need for operator involvement.

(2) When the UAV lacks the knowledge to handle the current task, horizontal adjustment of the autonomy level reduces the complexity of actions and allows operator intervention in the decision-making process. Corresponding limited-intervention methods of the manned aircraft are selected for different decision-making tasks, fully integrating the cognitive capabilities of the manned aircraft operator with the task execution capabilities of the UAV to achieve collaborative decision-making.

(3) For extreme situations such as communication interruption, the UAV first performs horizontal adjustment of the autonomy level, retaining a certain degree of autonomy while awaiting communication recovery. If the waiting time is exceeded, it vertically downgrades the autonomy level to execute lower-complexity actions for self-preservation such as returning to the base.

In summary, integrating variable UAV autonomy levels with limited manned aircraft intervention enables autonomous task execution in simple scenarios. This approach also effectively combines the operator's cognitive strengths with the UAV's mission capabilities in complex operational environments.

3.5.5.2 Analysis of Simulation Results for the Overall Task

From the perspective of the overall task process, the three decision-making tasks (i.e., obstacle avoidance, threat avoidance, and target engagement) are connected and combined. A comparative simulation analysis is conducted for three scenarios, including command-and-control, variable-autonomy control, and full-autonomy control of the UAV, as shown in Figure 3.24.

Scenario 2 for obstacle avoidance, threat avoidance, and attack decision-making is selected and connected to form the overall task process of the UAV. This scenario sequence exhibits two characteristics: first, under the command-control mode, the number of interactions increases significantly, failing to fully utilize the autonomy of the UAV, which imposes a heavier mental workload on the operator and results in resource wastage. Second, under the fully autonomous mode of the UAV, task completion cannot be guaranteed, leading to a sharp decline in task effectiveness or even an increased mission failure rate.

To facilitate comparative analysis, it is assumed that, under normal communication conditions, command control can complete all tasks within the scenario. Within this framework, the UAV's command-control

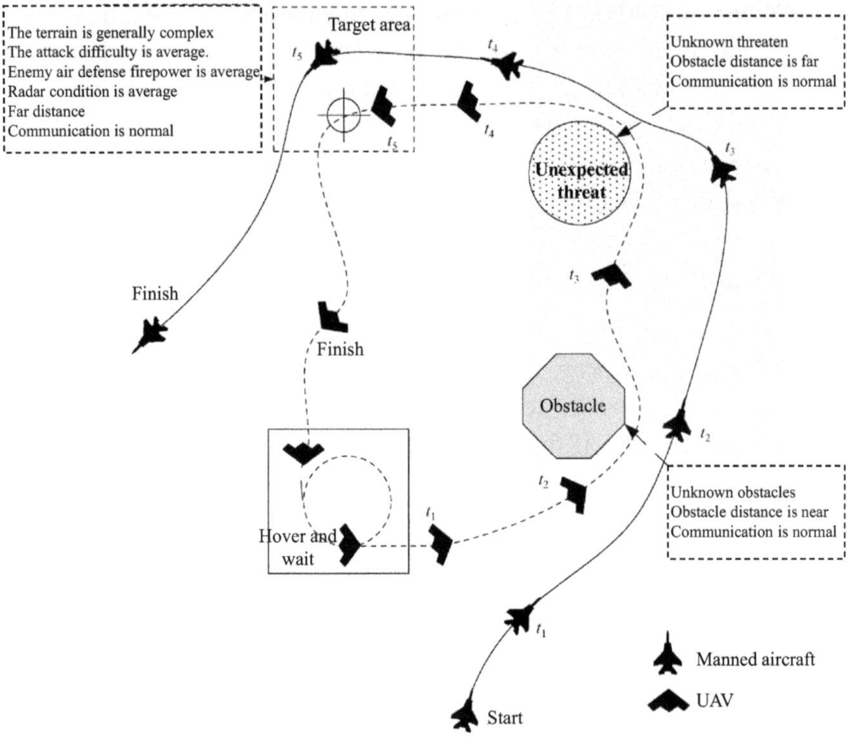

FIGURE 3.24 Scenario for the comprehensive process of UAV decision-making task

mode requires the highest number of interactions, while the full-autonomy mode requires the least.

3.5.5.2.1 Analysis of Task Complexity for UAV and Workload of Manned Aircraft Operator Figure 3.25 illustrates the task execution process of the UAV. Gray boxes represent the action sequences during each subtask process, while the positions of the boxes indicate the hierarchical level of the actions. The UAV remains under the approval management mode throughout the entire task process but undergoes certain fine-tuning within the same level. For example, encountering unknown obstacles causes the UAV to enter a hovering state and wait for instructions, thereby reducing the action hierarchy and autonomy level. After receiving intervention information from the manned aircraft, it selects obstacle avoidance, thereby increasing the autonomy level.

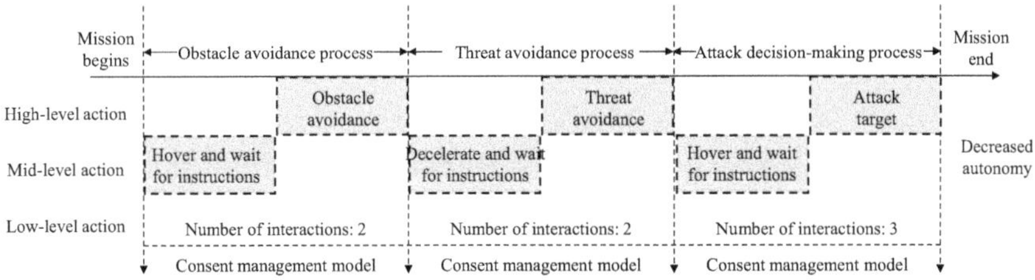

FIGURE 3.25 Task execution process of UAV

The task complexity of the UAV is calculated by Equation (3–25):

$$D(t) = 1 - \frac{1}{N_H + 0.5N_I} = 1 - \frac{1}{3 + 0.5 \times 3} = 0.78 \tag{3–25}$$

Assuming the UAV lacks autonomy, the task complexity when operations are completed via command control is defined as follows:

$$D(t) \approx 0 \tag{3–26}$$

When the UAV is fully autonomous and does not require operator assistance, the task complexity is defined by Equation (3–27) as follows:

$$D(t) \approx 1 \tag{3–27}$$

The higher the complexity of the action sequence executed by the UAV during the task process, the lower the task complexity faced by the operator. The task complexity of the operator is defined as follows:

$$\hat{D}(t) = 1 - D(t) \tag{3–28}$$

Thus, the task complexity of the operator under command control mode is as follows:

$$\hat{D}_1(t) = 1 - D_1(t) \approx 1 \tag{3–29}$$

The task complexity of the operator under variable autonomy mode is as follows:

$$\hat{D}_2(t) = 1 - D_2(t) = 1 - 0.78 = 0.22 \tag{3–30}$$

The task complexity of the operator under full autonomy mode is as follows:

$$\hat{D}_3(t) = 1 - D_3(t) \approx C \tag{3-31}$$

It can be observed that when the UAV operates with full autonomy, the task complexity faced by the operator is the lowest, approximately 0. Under the variable autonomy mode, the task complexity faced by the operator is intermediate at 0.22. Under the command control mode, the task complexity faced by the operator is the highest, approximately 1. The task complexity faced by the operator is positively correlated with the operator's mental workload; that is, the higher the task complexity is, the greater the operator's mental workload, and conversely, the lower the task complexity is, the smaller the workload.

3.5.5.2.2 Analysis of SA Capability of Manned Aircraft Operator It is assumed that the weighting of the impact on SA levels under various autonomy level modes is as shown in Table 3.21.

The SA level of the manned aircraft operator during the task process is calculated as follows:

Command control mode is through: $\mathrm{SA}_1 = \sum_{i=1}^{4} \frac{\mathrm{Ca}_i}{\mathrm{Cn}} \cdot \mathrm{wg}_1 = 0.25$

Variable autonomy mode is via: $\mathrm{SA}_2 = \sum_{i=1}^{4} \frac{\mathrm{Ca}_i}{\mathrm{Cn}} \cdot \mathrm{wg}_i = 0.5$

Full autonomy mode is through: $\mathrm{SA}_3 = \sum_{i=1}^{4} \frac{\mathrm{Ca}_i}{\mathrm{Cn}} \cdot \mathrm{wg}_i = 0.1$

TABLE 3.21 Weighting of Impact on SA Levels Under Various Autonomy Level Modes

Autonomy Level	Weight
Command control mode	0.25
Operator decision mode	0.15
Consent management model	0.5
Full autonomy mode	0.1

3.5.5.2.3 Analysis of Task Completion Degree for Manned Aircraft and UAV The task completion degree of the manned aircraft and UAV is calculated as follows:

Command control mode is through:

$$\text{Task completion rate} = \frac{n}{N} \times 100\% = 100\%$$

Variable autonomy mode is via:

$$\text{Task completion rate} = \frac{n}{N} \times 100\% = 100\%$$

Full autonomy mode is as follows: Task completion rate= $\frac{n}{N} \times 100\% = 0\%$,

where n is the number of completed subtasks, and N is the total number of subtasks.

It can be observed that although both the variable autonomy mode and the command control mode can ensure task completion, compared with the command control mode, the variable autonomy mode of the UAV results in lower task complexity for the operator, reduced operator pressure, and a higher SA level. Although the operator's task complexity is the lowest under the full autonomy mode, due to the almost nonexistent interaction between the UAV and the operator in this mode, the operator's SA level is minimized. When encountering situations beyond its knowledge scope, the UAV is almost incapable of making autonomous decisions.

To derive more general conclusions, the case scenarios included in the three decision-making tasks are combined and connected in series, generating $C_3^1 \times C_3^1 \times C_4^1 = 36$ distinct cases that essentially cover all possible situations during the UAV's mission scenario process. For each of these 36 cases, the aforementioned analysis is conducted, and the statistical values of the three metrics mentioned earlier are calculated.

As shown in Figure 3.26:

(1) Regarding the operator's task complexity, the command control mode exhibits the highest level, followed by the variable autonomy mode, while the full autonomy mode shows the lowest level. When the workload pressure on the manned aircraft operator is high, the mode can be switched from command control to variable autonomy.

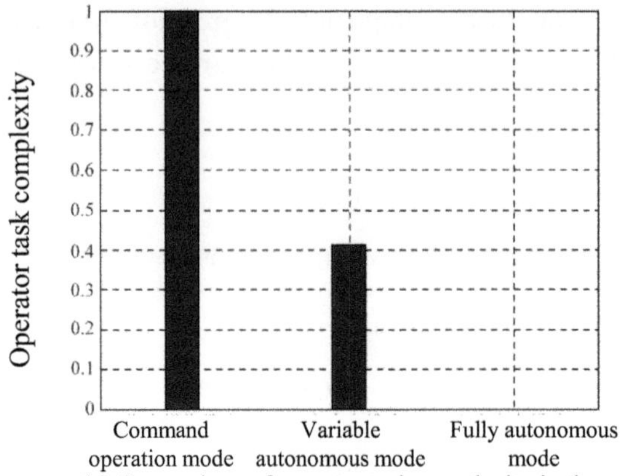

(a) Comparison of operator task complexity in the three modes

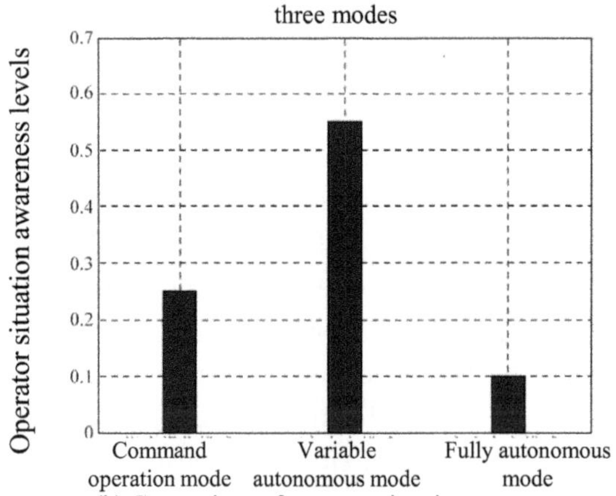

(b) Comparison of operator situation awareness levels under the three modes

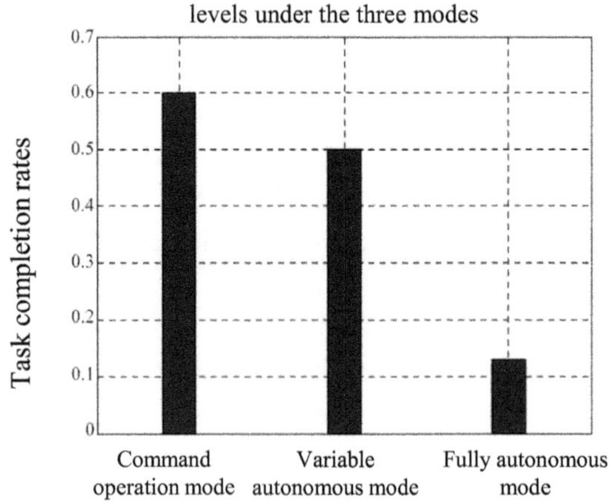

(c) Comparison of task completion rates under the

FIGURE 3.26 Comparison of operator task complexity, SA level, and task completion degree

(2) Regarding the operator's SA level of the battlefield, the variable autonomy mode yields the highest SA level, followed by the command control mode, while the full autonomy mode results in the lowest. Therefore, when the operator needs to maintain a high SA level, the mode should be switched to the variable autonomy mode.

(3) Regarding task completion degree, the command control mode achieves the highest task completion degree, as it is fully controlled by the operator. The variable autonomy mode combines UAV autonomy with manned aircraft intervention, resulting in an intermediate task completion degree. The full autonomy mode relies entirely on the autonomy level of the UAV to complete tasks, yielding the lowest task completion degree. Therefore, when the operator's mental workload is manageable, the command control mode should be prioritized for tasks requiring high precision.

By qualitatively representing the operator's task complexity, SA level, and task completion degree as distinct evaluation dimensions, as shown in Figure 3.27, the following conclusions can be drawn:

(1) Under the full autonomy mode, although the operator faces the lowest task complexity, it does not hold advantages in SA level or task completion degree. Therefore, when the UAV cannot guarantee autonomous handling of all possible situations, it should not operate in the full autonomy mode. This also indirectly reflects the significant potential and important value of enhancing the autonomous capabilities of UAVs.

(2) Under the command control mode, the task completion degree is the highest. Under normal communication conditions, UAVs can ensure task completion. However, this mode has a relatively high task complexity for the operator. As the operator devotes excessive attention to controlling the UAV, their SA level is compromised. Therefore, when environmental complexity is low and tasks are simple, the command control mode should be prioritized.

(3) Under the variable autonomy mode, the operator maintains a high SA level throughout the task process while facing moderate task complexity. Under normal communication conditions, this mode can ensure the completion of simple tasks such as obstacle avoidance and threat avoidance, as well as less complex attack decision-making tasks. During communication interruptions, the UAV can fully utilize its autonomous capabilities to ensure its own safety while awaiting communication recovery, as shown in Figure 3.27.

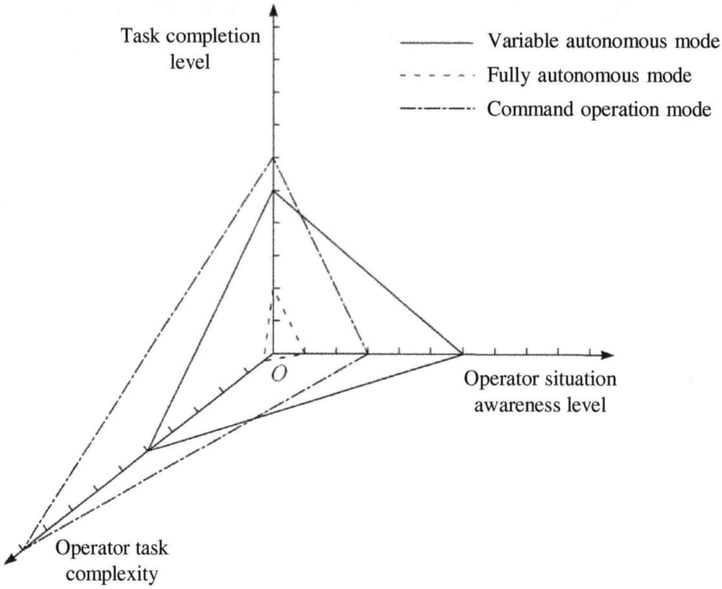

FIGURE 3.27 Comprehensive comparison of three UAV control modes

In summary, the manned aircraft and UAV limited-intervention collaborative decision-making mechanism can leverage advantages in reducing operator task complexity and improving SA levels. The UAV's variable autonomy adjustment mechanism enables full integration of human cognitive strengths with UAV mission execution capabilities. To implement this collaborative decision-making model, the following recommendations are proposed:

(1) The manned aircraft and UAV limited-intervention collaborative decision-making model enables the operator to maintain task complexity at a moderate level while achieving high SA. The task completion degree remains comparable to that of the command control mode, demonstrating significant application prospects and development potential.

(2) Under the limited-intervention collaborative decision-making mode, the key factor affecting the task completion degree of the UAV is the communication status. This highlights the importance of enhancing

the autonomous mission execution capability of the UAV under the communication interruption scenarios.

(3) Given the limited autonomy level of the UAV, it cannot yet fully autonomously execute missions. The command control mode can serve as a reliable backup solution when task requirements are high, or the operator's mental workload pressure is manageable.

3.6 CHAPTER SUMMARY

Through the description of the limited-intervention collaborative decision-making tasks between manned aircraft and UAV, the establishment of collaborative obstacle avoidance, threat avoidance, and attack decision-making models was carried out, along with corresponding simulation case studies. The main work completed in this chapter includes the following:

(1) Based on the definition of autonomy levels, the actions for UAV obstacle avoidance, threat avoidance, and attack decision-making were categorized.

(2) Using FCM theory, FCM models for UAV obstacle avoidance, threat avoidance, and attack decision-making were established, respectively.

(3) Guided by the limited-intervention mechanism, decision-making models for obstacle avoidance, threat avoidance, and collaborative attack have been developed for manned aircraft.

(4) Through case simulations, the limited-intervention collaborative obstacle avoidance, threat avoidance, and attack decision-making models for manned aircraft and UAV were analyzed, discussed, and summarized.

CHAPTER 4

Analysis of Cognitive and Interactive Cooperative Decision-Making Mechanisms between Manned Aircraft and UAVs

THIS CHAPTER PROPOSES A framework for the cognitive and interactive cooperative decision-making mechanism between manned aircraft and UAVs. It also analyzes the primary methods and pros and cons of cognitive load assessment methods for manned aircraft pilots and further defines and establishes the conceptual framework and modeling approach for intelligent affective modes in UAVs.

4.1 FRAMEWORK OF THE COGNITIVE AND INTERACTIVE COOPERATIVE DECISION-MAKING MECHANISM

The framework for the cognitive and interactive cooperative decision-making mechanism between manned aircraft and UAVs consists of three layers, including the SA layer, the decision-making and reasoning layer,

DOI: 10.1201/9781003791713-4

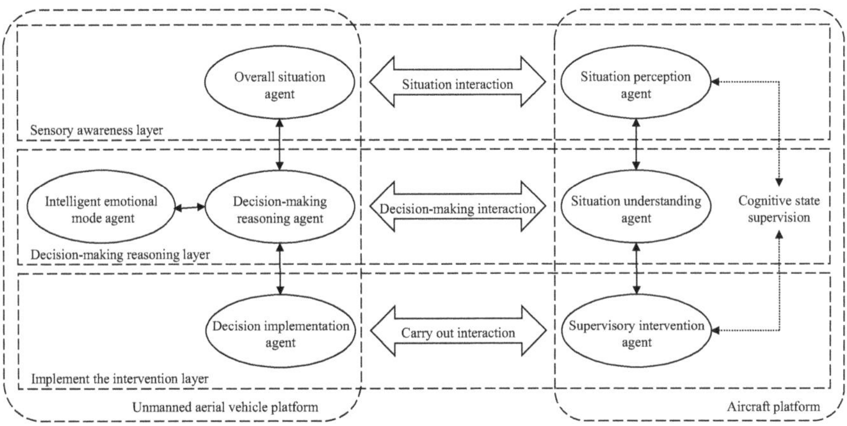

FIGURE 4.1 Framework of the cognitive and interactive cooperative decision-making mechanism between manned aircraft and UAVs

and the execution and intervention layer, as illustrated in Figure 4.1. The SA layer primarily involves information collection, perception, and integration by both the manned aircraft pilot and the UAVs before cooperative decision-making. The decision-making and reasoning layer involves UAVs performing intelligent decision-making and reasoning based on integrated situational awareness and the pilot's comprehension of the current situation. The execution and intervention layer includes the execution of decisions by the UAV and its response to supervision and intervention from the manned aircraft. Across all three layers, decision-making is primarily led by UAVs, supplemented by the manned aircraft, with continuous monitoring of the pilot's cognitive state, reflecting the interactive nature of cognitive intelligence.

4.1.1 SA Layer

The situational awareness (SA) layer is the part of the framework that directly interacts with the external environment. The UAV collects external environmental information, its own status, and the pilot's cognitive load through sensors, integrating these into the comprehensive SA module. The manned aircraft pilot perceives the current situational elements under the guidance of their attention. The UAV feeds integrated situational information back to the pilot, further enhancing the pilot's understanding

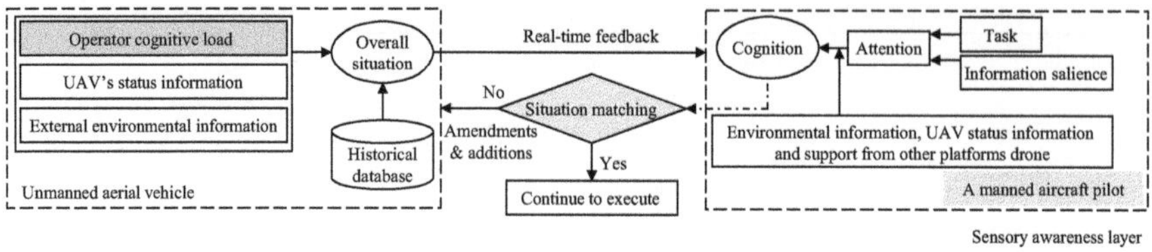

FIGURE 4.2 Implementation process of the SA layer

of the current situation of their surroundings. The pilot can supplement or modify the information acquired by the UAV. The implementation process is shown in Figure 4.2.

Unlike the limited-intervention cooperative decision-making mechanisms, the SA layer requires focused modeling of the cognitive load of the manned aircraft pilot. This enables the UAV to perceive the pilot's mental state in addition to the external environment and its own status, thereby actively adapting to and cooperating with the pilot in intelligent decision-making to improve the efficiency of the human-machine cooperative loop.

4.1.2 Decision-Making and Reasoning Layer

The decision-making and reasoning layer is the middle layer of the cognitive and interactive cooperative decision-making framework. It primarily receives preprocessed and integrated situational information from the SA layer to perform intelligent decision-making and reasoning. Initially, the UAV determines the intelligent emotional mode that is suitable for the current situation, thereby autonomously adjusting its decision-making preferences. The manned aircraft pilot can then interact with the UAVs based on their understanding of the situation. The implementation process of the model is shown in Figure 4.3.

In the decision-making and reasoning layer, the UAV's intelligent affective mode and decision-making reasoning mechanism form the core of the cognitive-interactive cooperative framework. The intelligent emotional mode drives the UAV's decision-making reasoning, with different modes producing different decision preferences. The UAV's intelligent emotional mode is influenced by the external environment, its own status, and the pilot's cognitive load. If the external environment is dynamic and complex, the UAV's status is poor, and the pilot's cognitive load is high, the decision-making reasoning of the UAV should lean toward conservatism to reduce the pilot's mental workload and ensure their safety. During the

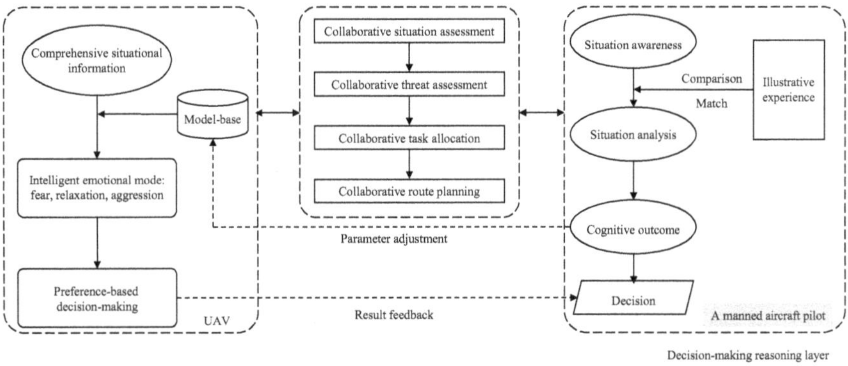

FIGURE 4.3 Implementation process of the decision-making and reasoning layer

process of the intelligent emotional mode and decision-making reasoning, the manned aircraft pilot subjectively adjusts parameters of the decision-making reasoning model based on their cognitive results. Simultaneously, the UAV provides real-time feedback on decision results to the pilot.

4.1.3 Execution and Intervention Layer

In the execution and intervention layer, the manned aircraft pilot determines whether to intervene based on the alignment between the UAV's intelligent decision outcomes and their own cognitive judgment. If the results align with the pilot's judgment, then the decisions are executed. If a significant discrepancy exists between the intelligent decision and the pilot's assessment, cooperative decision-making is then initiated. When the pilot is in a promising cognitive condition, four intervention methods can be adopted, including environmental intervention, knowledge intervention, decision intervention, and manual control, with increasing levels of intervention intensity. Environmental intervention refers to the pilot influencing or modifying the operational environment of the UAV. Knowledge intervention involves the pilot supplementing or modifying the UAV's knowledge base. Decision intervention involves the pilot modifying the decision results of the UAV. Manual control refers to the pilot controlling and making decisions for the UAV by setting waypoints or assigning tasks. When the pilot's cognitive state is poor and they are unable to intervene, the UAV can adjust its intelligent operation mode and reinitiate decision-making reasoning. The implementation process of this model is shown in Figure 4.4.

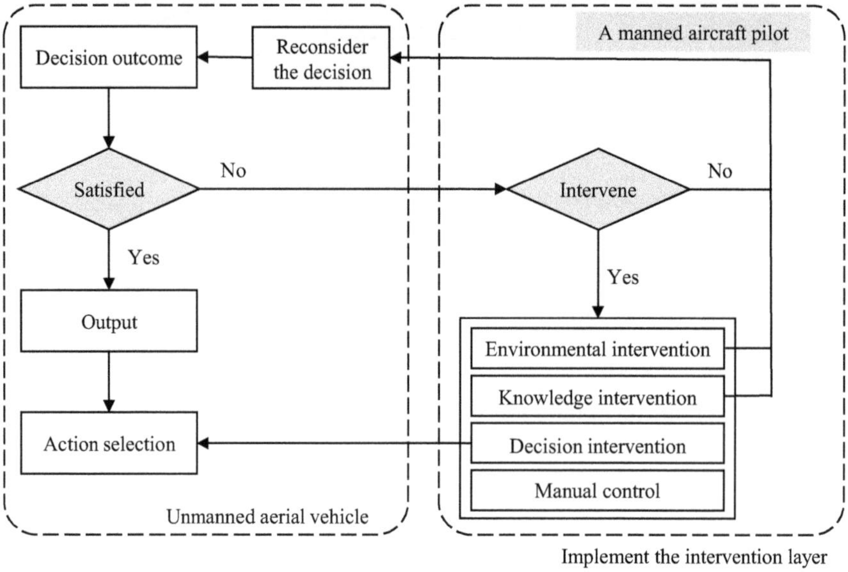

FIGURE 4.4 Implementation process of the execution and intervention layer

In the cognitive and interactive cooperative decision-making mechanism, integrating the pilot's cognitive load and the UAV's intelligent emotional mode into the SA, decision-making and reasoning, and execution and intervention layers is crucial for constructing an effective model.

4.2 COGNITIVE LOAD ASSESSMENT FOR MANNED AIRCRAFT PILOTS

Ensuring that manned aircraft pilots can perform human-machine cooperative tasks efficiently and accurately without increasing their task execution load is a common concern in the field of human-machine cooperation (HMC) [67]. Human factors and ergonomics (HFE), as a discipline of HMC, have received widespread attention globally [68]. According to the International Ergonomics Association, HFE is a scientific field that studies the interactions between humans and systems or elements, applying theories, principles, data, and methods to design systems that enhance human-machine system performance. There are three main research areas, as shown in Table 4.1 [69, 70]. The focus of HFE lies in system design to adapt to human needs, capabilities, and limitations, with the key to system design being the study and understanding of human states [71, 72]. In the context of cooperative decision-making between manned aircraft

TABLE 4.1 Three Main Research Areas of HFE

Research Area	Specific Explanation
Physical ergonomics	Related to human physical activities, focusing on human physical characteristics.
Cognitive ergonomics	Related to psychological characteristics, focusing on human cognitive features.
Macro ergonomics	Related to sociotechnical systems, focusing on human psychosocial characteristics.

and UAVs, manual operation and cognitive states, as key factors affecting human perception, decision-making, and action, are fundamental issues that need to be addressed.

4.2.1 Common Evaluation Methods

Workload generally refers to the effort expended by a pilot to maintain task performance, where it leans toward the efforts made physically and physiologically [73]. In human-machine cooperative tasks, workload manifests as the physical burden on the manned aircraft pilot due to prolonged task duration or high operational intensity. Due to the multidimensional characteristics of cognitive workload, commonly used indicators in existing studies, such as information processing capacity, time occupancy, and attentional resources, are relatively abstract and suitable for conceptual analysis but challenging to apply in practical assessment and modeling. From the perspective of task difficulty, workload can be considered to follow a "U"-shaped curve as task difficulty increases [74], as shown in Figure 4.5. Based on the relationship between workload, performance level, and task difficulty, it can be divided into six different regions.

In region A, as task difficulty increases, workload gradually rises from a high-level, ultra-low arousal state and then slowly declines, while performance increases rapidly. In region B1, as task difficulty increases, workload decreases rapidly, and performance level slowly rises to an optimal state. In region B2, as task difficulty increases, workload remains at the lowest level, and performance level remains optimal. In region B3, as task difficulty increases, workload rises rapidly, and performance level slowly declines. In region C, as task difficulty increases, workload gradually rises, and performance level drops rapidly. In region D, as task difficulty increases, workload enters an ultra-high state, and performance level reaches an ultra-low level. Analysis of these six regions shows that both

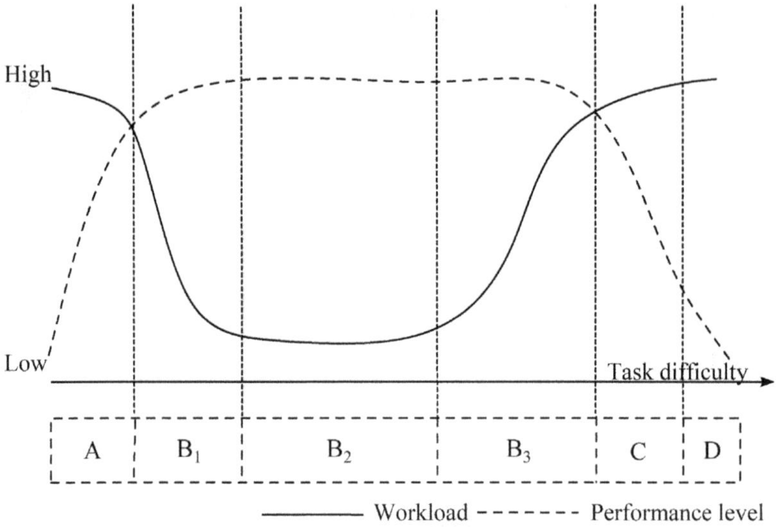

FIGURE 4.5 Relationship between workload, task difficulty, and performance level

excessively high and low workloads are detrimental to task completion. Only by maintaining workload at an appropriate level can the pilot's performance level reach its optimal state. Therefore, in workload assessment research, different task difficulties can be designed to simulate varying levels of workload experienced by subjects.

Cognitive load generally refers to the total mental energy required for mental activities in working memory to complete a task, also defined as the effort that tends to be made mentally and cognitively [75]. In human-machine cooperative tasks, it manifests as the mental burden on the manned aircraft pilot when processing a large number of tasks in a short time. Cognitive load theory suggests that limited working memory is the "bottleneck" in human cognitive processes, as well as the main source of cognitive load, necessitating in-depth exploration of human cognition and memory processes.

Figure 4.6 depicts the human memory model, describing human memory and cognitive processes, defining three types of memory:

(1) Sensory memory: The first step in human cognitive processes is acquiring external information through the senses. Sensory memory stores vast amounts of sensory information such as text, images

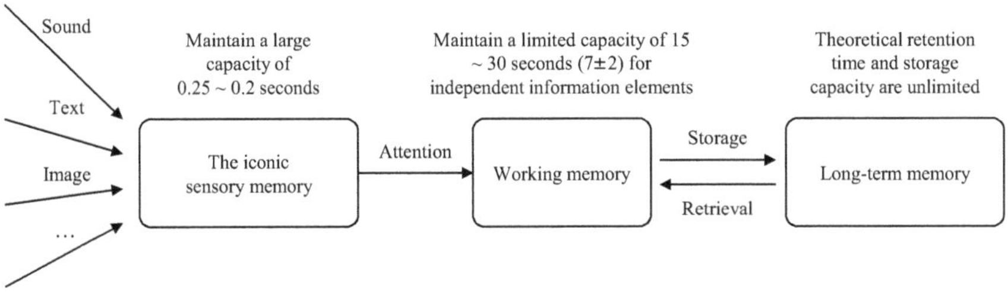

FIGURE 4.6 Human memory model

(visual), and sounds (auditory), but this information is retained only briefly, around 0.25 to 2 seconds, with most of it not entering working memory. However, when attention is paid to the sensory memory information, this information is preserved and enters working memory.

(2) Working memory: Actively processes sensory information, like organizing and comparing. Working memory has a limited capacity, typically of around (7 ± 2) independent information elements [76]. Once the information elements exceed the capacity of working memory, cognitive load occurs.

(3) Long-term memory: Stores information processed by working memory, including memories, knowledge, and experiences. The storage capacity and retention time of long-term memory are theoretically unlimited.

Table 4.2 summarizes common workload and cognitive load evaluation methods, accompanied by a corresponding analytical overview.

(1) Compared to subjective and performance-based evaluation methods, physiological evaluation methods are more suitable for assessing workload and cognitive load due to their objectivity, real-time capability, and continuity. When a single physiological indicator cannot fully assess workload/cognitive load levels, multiple physiological indicators can be combined for evaluation by using comprehensive evaluation methods.

(2) Among physiological evaluation methods, eye movement metrics are less intrusive than EEG and multiparameter physiological metrics.

TABLE 4.2 Summary of Current Common Workload and Cognitive Load Evaluation Methods

Type	Definition	Methods	Advantages	Disadvantages
Subjective evaluation	Subjects evaluate their performance based on personal feelings after task completion.	NASA-TLX* Scale, SWAT** Scale, MCH*** Scale	(1) Direct assessment of workload and cognitive load without instruments, with simple operation and statistical analysis. (2) Directly reflects subjects' mental workload, highly sensitive to changes in mental load levels.	(1) Lacks real-time capability as statistics and calculations are performed post-test. (2) High individual variability, and subjective evaluation is closely related to factors such as personality, action strategies, physiological/psychological state, and mood.
Performance-based evaluation	A secondary task is added while the subject performs the primary task, and it indirectly affects each one's performance.	Secondary tasks: arithmetic task, memory tasks, visual search. Metrics: reaction time, accuracy rate.	(1) Addresses issues where primary task performance is hard to measure and is well grounded theoretically. (2) High sensitivity in measuring mental workload or cognitive load.	(1) Intrusive, mainly interfering with the primary task and not suitable for real-world environments. (2) Lacks real-time capability.
Physiological evaluation	Assesses subjects' mental state by measuring various physiological-based indicators.	Eye movement: gaze, saccade, blink rate, pupil area. EEG: frequency bands, event-related potentials, fMRI.	(1) Offers objectivity, real-time capability, and continuity. (2) Advances in integrated, intelligent, automated physiological measurement reduce intrusiveness and simplify data analysis.	Suffers from specificity issues in sensitivity. For example, a particular physiological indicator may be sensitive only to certain types of tasks and not others.

		Multiparameter physiological metrics: heart rate, electromyography, blood oxygen concentration, respiratory rate.	(3) Physiological measurement can process data from different time periods simultaneously and perform analyses during continuous data collection.	
Comprehensive Evaluation	Combines multiple methods instead of relying on a single method or indicator.	Eye movement metrics + EEG metrics + multiparameter physiological metrics.	(1) More reliable by using multiple methods or indicators instead of relying on any one of them. (2) Suitable for different contexts and outputs due to the multidimensional nature of workload/cognitive load.	Synchronization issues may arise with multiple methods or indicators.

* *NASA-TLX:* National Aeronautics and Space Administration Task Load Index. A multidimensional assessment scale consisting of six dimensions (i.e., mental demand, physical demand, temporal demand, performance, effort, and frustration), widely used in flight research and related fields.

** *SWAT:* Subject Workload Assessment Technique. A scale composed of three dimensions (i.e., time load, mental effort load, and psychological stress load), each divided into high, medium, and low levels, combined into a single measurement indicator.

*** *MCH:* Modified Cooper-Harper Scale. A scale that divides flight operation levels into 10 grades, each with corresponding definitions. After task completion, pilots rate the scale based on their subjective feelings to evaluate their workload/cognitive load.

Eye trackers (e.g., camera-, desktop-, and glasses-type) can collect eye movement data, measure subjects' attention, and analyze human cognitive processes. Therefore, eye movement metrics should be the primary measurement indicators for assessing workload and cognitive load.

(3) Referring to the definition of workload, it leans toward the physical and physiological efforts being made. Therefore, when selecting indicators for workload assessment, multiparameter physiological indicators that reflect the subject's physiological state should be chosen, combined with eye movement metrics to form comprehensive workload assessment indicators.

(4) Referring to the definition of cognitive load, it leans toward the efforts being made mentally and cognitively. Therefore, when selecting indicators for cognitive load assessment, EEG indicators that better reflect the subject's cognitive state should be chosen, combined with eye movement metrics to form comprehensive cognitive load assessment indicators.

4.2.2 Classification Principle Based on SVM

Support vector machines (SVM) are a commonly used method for addressing the mental workload and cognitive load evaluation problem. SVM is based on the structural risk minimization theory, constructing an optimal hyperplane in the feature space to achieve global optimization of the model. For linearly separable binary classification problems, SVM reduces computational complexity by finding support vectors, seeking an optimal separation interface between two classes of sample sets to maximize the margin between the hyperplane and sample points. For nonlinearly separable binary classification problems, SVM uses kernel functions to map nonlinearly separable low-dimensional feature vector spaces into linearly separable high-dimensional feature vector spaces, simplifying inner product operations in the mapped space and avoiding direct computation in the low-dimensional space. Then, it uses linearly separable SVM for classification, enabling linear analysis of nonlinear features of samples in the high-dimensional feature vector space. The principle of SVM is briefly described as follows:

Given a sample set $D=\{(x_1,y_1),(x_2,y_2),\cdots,(x_n,y_n)\}$ with its labels $y_i\in\{-1,+1\}$, the dividing hyperplane in the sample space can be described through the linear equation:

$$\omega^{\mathrm{T}}x+b=0 \tag{4-1}$$

Here, $\omega=(\omega_1,\omega_2,\cdots,\omega_d)$ is the normal vector determining the direction of the hyperplane, and b is the displacement term determining the distance between the hyperplane and the origin. If the hyperplane can correctly classify the training samples, then:

$$\begin{cases}\omega^{\mathrm{T}}x_i+b\geq+1, & y_i=+1\\ \omega^{\mathrm{T}}x_i+b\leq-1, & y_i=-1\end{cases} \tag{4-2}$$

For SVM, the hyperplane with the maximum margin needs to be found; that is, the sum of the distances from two heterogeneous support vectors to the hyperplane $\frac{2}{\|\omega\|}$ is maximized:

$$\begin{cases}\max_{\omega,b}\dfrac{2}{\|\omega\|}\\ \text{s.t. } y_i(\omega^{\mathrm{T}}x_i+b)\geq 1, \quad i=1,2,\cdots,n\end{cases} \tag{4-3}$$

Clearly, to maximize the margin, it is only necessary to maximize $\|\omega\|^{-1}$, which is equivalent to minimizing $\|\omega\|^2$. Thus, Equation (4–3) can be rewritten as follows:

$$\begin{cases}\min_{\omega,b}\dfrac{1}{2}\|\omega\|^2\\ \text{s.t. } y_i(\omega^{\mathrm{T}}x_i+b)\geq 1, \quad i=1,2,\cdots,n\end{cases} \tag{4-4}$$

By adding Lagrange multipliers $\alpha_i\geq 0$ to each constraint of Equation (4.4), the dual problem of Equation (4.5) is obtained:

$$\begin{cases}\max_{\alpha}\left[\sum_{i=1}^{n}\alpha_i-\dfrac{1}{2}\sum_{i=1}^{n}\sum_{j=1}^{n}\alpha_i\alpha_j y_i y_j x_i^{\mathrm{T}}x_j\right]\\ \text{s.t. } \sum_{i=1}^{n}\alpha_i y_i=0\\ \alpha_i\geq 0, \quad i=1,2,\cdots,n\end{cases} \tag{4-5}$$

The Lagrange multiplier α_i corresponds to the training sample (x_i,y_i), and Equation (4.4) has inequality constraints. Therefore, the aforementioned process must satisfy the KKT (Karush-Kuhn-Tucker) conditions, which are described as follows:

$$\begin{cases} \alpha_i \geq 0 \\ y_i f(x_i) - 1 \geq 0 \\ \alpha_i (y_i f(x_i) - 1) = 0 \end{cases} \tag{4-6}$$

After solving for α, ω and b can be obtained to form the model.

$$f(x) = \omega^{\mathrm{T}} x + b = \sum_{i=1}^{n} \alpha_i y_i x_i^{\mathrm{T}} x + b \tag{4-7}$$

4.3 INTELLIGENT EMOTIONAL MODE OF UAVS

The intelligent affective mode of a UAV refers to its ability to exhibit human-like emotional characteristics in autonomous behavior. Based on situational assessment and the pilot's cognitive state, it selects an appropriate affective mode to influence the UAV's autonomous decision-making, reasoning, and cooperative task execution. The overall decision flowchart is shown in Figure 4.7.

4.3.1 Definition of UAV Intelligent Emotional Mode

Cognitive psychology research suggests that human decision-making relies not only on past experiential knowledge and current

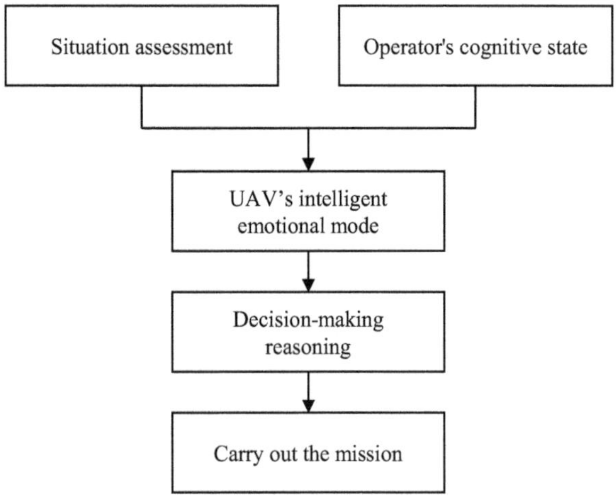

FIGURE 4.7 Decision flowchart of the intelligent emotional mode of UAVs

decision-making information but also on current emotional states. To achieve a higher level of autonomy in UAVs, it is necessary to propose and establish an intelligent emotional mode for UAVs that mimics human decision-making.

The environment plays a significant role in traditional UAV decision-making. UAVs process collected information and make corresponding behavioral choices based on dynamic changes in the external environment. However, this process does not consider internal UAV information. For example, the platform status and system requirements of the UAV are often assumed to be in an ideal state [77]. In complex and volatile battlefield environments, factors such as the UAV's equipment status and remaining fuel are critical constraints on UAV operations and must be considered during decision-making.

The UAV intelligent emotional mode can be divided into three types: fear, relaxed, and aggressive. Under these three intelligent emotional modes, the UAV will have different decision preferences for the same task.

Fear: When the UAV is in a fearful state, it tends to make decisions that are easier to accomplish, have smaller benefits, but ensure self-preservation.

Relaxed: When the UAV is in a relaxed state, it tends to make decisions with moderate difficulty and average benefits.

Aggressive: When the UAV is in an aggressive state, it tends to make decisions that are difficult to accomplish but offer high benefits.

The UAV adjusts its emotional state based on the complex and changing battlefield environment, its own status, and the pilot's cognitive load. The UAV feeds its emotional state back into its decision-making system, resulting in different decision preferences under different emotional states. When the battlefield environment is complex, the UAV's own status is poor, and the pilot's cognitive load is high, the UAV's intelligent emotional mode should be in a "fear" mode. In this state, decision-making reasoning prioritizes ensuring the UAV's safety. When the UAV's own status is good, communication with the manned aircraft is normal, and the pilot's cognitive load is low, the UAV's intelligent emotional mode should be in an "aggressive" mode. In this state, decision-making reasoning prioritizes the mission benefits of the UAV.

4.3.2 FCM Model of UAV Intelligent Emotional Mode

This subsection establishes the FCM model (i.e., Figure 4.8) for the UAV intelligent emotional mode by selecting key conceptual nodes that

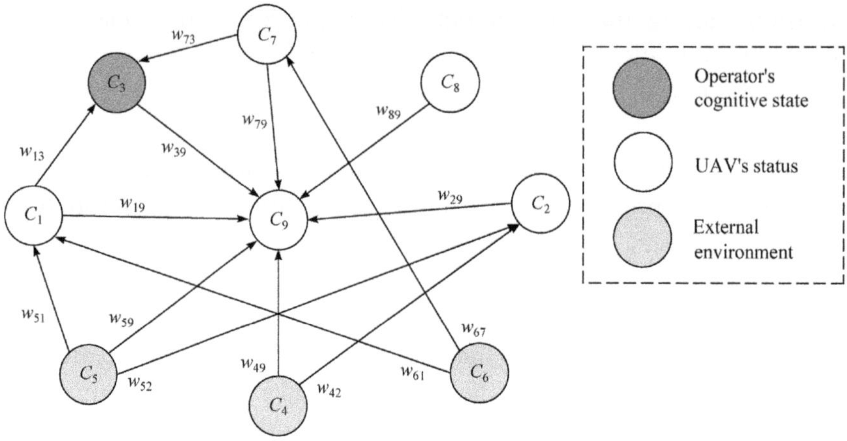

FIGURE 4.8 FCM model of UAV intelligent emotional mode

TABLE 4.3 Definitions of FCM Conceptual Nodes in the UAV Intelligent Emotional Mode

Node	Conceptual Node Name
C_1	Manned-UAV Communication Quality
C_2	UAV Fuel Remaining
C_3	Manned Aircraft Pilot Cognitive Load
C_4	Terrain Complexity
C_5	Atmospheric Density of Environment
C_6	Relative Humidity of Environment
C_7	UAV Radar Detection Capability
C_8	Distance between UAV and Threat Point
C_9	UAV Intelligent Emotional Mode

significantly influence it. The definitions of the conceptual nodes in the UAV intelligent emotional mode FCM model are presented in Table 4.3.

4.3.2.1 Manned-UAV Communication Quality

The quality of communication between the manned and unmanned vehicles is crucial for UAV decision-making and also affects the pilot's cognitive load. Its normalization function is as follows:

$$C_1 = \tanh\left(\frac{1}{n_r} \times \frac{\sum_{y=1}^{n_r} (M_r)_y}{\sum_{y=1}^{n_t} (M_t)_x}\right) \tag{4–8}$$

Here, n_r is the number of packets received, n_t is the number of packets sent, M_r is the size of sent information, and M_t is the size of received information [78].

4.3.2.2 UAV Fuel Remaining

The normalization function for UAV fuel remaining is as follows:

$$C_2 = \frac{V_{\text{left}}}{V_{\text{max}}} \tag{4–9}$$

V_{left} is the remaining fuel amount of the UAV, and V_{max} is the maximum fuel capacity of the UAV.

4.3.2.3 Manned Aircraft Pilot Cognitive Load

The cognitive load of the manned aircraft pilot is processed and calculated through the SVM-based cognitive load assessment model.

4.3.2.4 Terrain Complexity

Different terrains affect the fuel required for UAV flight. Terrain complexity, ranked from high to low, includes mountains, plateaus, hills, basins, plains, among others, and it can be fuzzified to the range of [0,1].

4.3.2.5 Atmospheric Density of Environment

Higher atmospheric density increases the resistance on the UAV, requiring more fuel, and its normalization function is as follows, where ρ is the atmospheric density:

$$C_5 = \begin{cases} 1, & \rho \geq 1.225 \\ \dfrac{\rho - 0.104}{1.225 - 0.104}, & \rho < 1.225 \end{cases} \tag{4–10}$$

4.3.2.6 Relative Humidity of Environment

Higher relative humidity increases the resistance on the UAV and affects electromagnetic wave propagation. Its normalization function is as follows.

$$C_6 = \frac{d_1}{d_2} \tag{4–11}$$

d_1 is the water vapor density per unit volume, and d_2 is the saturated water vapor density.

4.3.2.7 UAV Radar Detection Capability

The normalization function for UAV radar detection capability is as follows:

$$C_7 = \tanh\left(\frac{R_{SN}}{D_0}\right) \tag{4–12}$$

Here, R_{SN} is the maximum signal-to-noise ratio during UAV radar operation, and D_0 is the minimum detectable signal-to-noise ratio.

4.3.2.8 Distance between UAV and Threat Point

The normalization function for the distance between the UAV and the threat points is as follows:

$$C_8 = k \cdot \sum_{i=1}^{n} \frac{c_i}{d_i} \tag{4–13}$$

where k is a normalization parameter, c_i is the weight of the i-th threat point, and d_i is the distance between the UAV and the i-th threat point.

4.3.2.9 UAV Intelligent Emotional Mode

The conceptual node for the UAV intelligent emotional mode is the output node of the model, inferred from other nodes. The value range of the UAV intelligent emotional mode can be determined based on expert knowledge. For example, the fear mode ranges between [0, 0.4], the relaxed mode ranges from (0.4, 0.7], and the aggressive mode ranges from (0.7, 1].

Explanations of causal relationships for some nodes in the UAV intelligent emotional mode FCM model are provided in Table 4.4.

The weight matrix of the UAV intelligent emotional mode FCM model is as follows:

$$w = \begin{matrix} 0 & 0 & w_{13} & 0 & 0 & 0 & 0 & 0 & w_{19} \\ 0 & 0 & 0 & 0 & 0 & 0 & 0 & 0 & w_{29} \\ 0 & 0 & 0 & 0 & 0 & 0 & 0 & 0 & w_{39} \\ 0 & w_{42} & 0 & 0 & 0 & 0 & 0 & 0 & w_{49} \\ w_{51} & w_{52} & 0 & 0 & 0 & 0 & 0 & 0 & w_{59} \\ w_{61} & w_{62} & 0 & 0 & 0 & 0 & w_{67} & 0 & w_{69} \\ 0 & 0 & w_{73} & 0 & 0 & 0 & 0 & 0 & w_{79} \\ 0 & 0 & 0 & 0 & 0 & 0 & 0 & 0 & w_{89} \\ 0 & 0 & 0 & 0 & 0 & 0 & 0 & 0 & 0 \end{matrix} \tag{4–14}$$

TABLE 4.4 Explanations of Node Causal Relationships

Node Causal Relationships	**Interpretations of the Causal Relationship**
$C_1 \xrightarrow{w_{13}} C_3$	High communication quality between manned and UAVs allows the manned vehicle to easily obtain UAV information, reducing the pilot's cognitive load. Therefore, C_3 has a negative influence on C_1.
$C_4 \xrightarrow{w_{42}} C_2$	More complex terrain requires more fuel for the UAV. Therefore, C_2 has a negative influence on C_4.
$C_6 \xrightarrow{w_{62}} C_2$	Increased relative humidity increases UAV flight resistance and fuel consumption. Therefore, C_2 has a negative influence on C_6.
$C_8 \xrightarrow{w_{89}} C_9$	According to the definition of the UAV intelligent emotional mode, the further the UAV is from a threat point, the more relaxed its emotional state, and the larger the value of C_9. Therefore, C_9 has a positive influence on C_8.
$C_1 \xrightarrow{w_{19}} C_9$	Better communication quality between humans and machines makes the UAV more inclined toward an aggressive intelligent emotional mode. Therefore, C_9 has a positive influence on C_1.
$C_3 \xrightarrow{w_{39}} C_9$	A higher cognitive load of the manned aircraft pilot makes the UAV more inclined toward a fear mode. Therefore, C_9 has a negative influence on C_3.

The values of the adjacency weight matrix can be defined by experts using semantic rules. The IF-THEN rule form is as follows:

IF the cause node C_i changes (none, small, medium, large, very large);

THEN the effect node C_j changes accordingly (none, small, medium, large, very large).

Assuming there is an influencing relationship set T from C_i to C_j, the degree of relationship is represented by a semantic variable μ with a value range of [−1, 1]. The influencing relationship set for the 9-category semantic variables is expressed as follows.

$$T(\text{influence}) = \left\{ \begin{array}{l} \text{negatively very strong}(\text{nvs}), \text{ negatively stong}(\text{ns}), \\ \text{negatively medium}(\text{nm}), \\ \text{negatively weak}(\text{nw}), \text{zero}(\text{z}), \\ \text{positively weak}(\text{pw}), \text{ posiively medium}(\text{pm}), \\ \text{positively strong}(\text{ps}), \text{ positively very strong}(\text{pvs}) \end{array} \right\} \tag{4–15}$$

Typically, the semantic relationships between nodes are determined using expert methods. The values of the semantic variables are determined using the centroid method through collective summation and are used as the corresponding weights w_{ij} in the adjacency matrix. A schematic diagram of the semantic variable values is shown in Figure 4.9.

The adjacency weight matrix of the UAV intelligent emotional mode FCM model is as follows:

$$w = \begin{array}{ccccccccc} 0 & 0 & 0.25 & 0 & 0 & 0 & 0 & 0 & 0.75 \\ 0 & 0 & 0 & 0 & 0 & 0 & 0 & 0 & 0.75 \\ 0 & 0 & 0 & 0 & 0 & 0 & 0 & 0 & 0.75 \\ 0 & 0.25 & 0 & 0 & 0 & 0 & 0 & 0 & 0.5 \\ -0.5 & -0.25 & 0 & 0 & 0 & 0 & 0 & 0 & -0.75 \\ -0.25 & -0.25 & 0 & 0 & 0 & 0 & -0.5 & 0 & 0 \\ 0 & 0 & 0.25 & 0 & 0 & 0 & 0 & 0 & 0.5 \\ 0 & 0 & 0 & 0 & 0 & 0 & 0 & 0 & 0.25 \\ 0 & 0 & 0 & 0 & 0 & 0 & 0 & 0 & 0 \end{array} \tag{4–16}$$

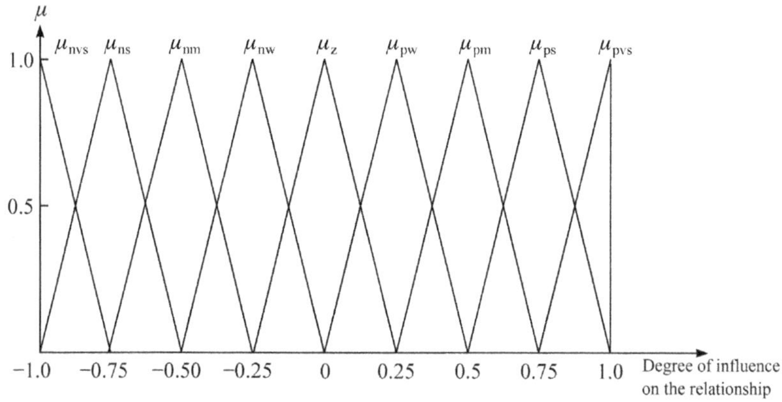

FIGURE 4.9 Schematic diagram of semantic variable values

TABLE 4.5 Node State Values of the UAV Intelligent Emotional Mode FCM Model

Node	Group 1	Group 2	Group 3	Group 4
C_1	0.8	0.4	0.2	0
C_2	0.8	0.6	0.3	0.8
C_3	0.3	0.5	0.8	0.9
C_4	0.25	0.75	0.5	0.5
C_5	0.3	0.7	0.8	0.3
C_6	0.4	0.5	0.6	0.4
C_7	0.9	0.4	0.23	0.9
C_8	0.5	0.6	0.8	0.8

4.3.3 Case Analysis

Table 4.5 provides the node state values for the UAV intelligent emotional mode FCM model.

Group 1: Communication quality between the manned and unmanned vehicles is good, the UAV has sufficient fuel and strong detection capability, the natural environment is suitable for UAV flight, and the pilot's cognitive load is low. Based on subjective judgment, the UAV's intelligent emotional mode should be in an aggressive state.

Group 2: All nodes of the model are at average levels. Based on subjective judgment, the UAV's intelligent emotional mode should be in a relaxed state.

Group 3: Communication quality between the manned and unmanned vehicles is poor. The UAV has insufficient fuel and weak radar detection capability, the terrain is relatively harsh, and the pilot's cognitive load is high. Based on subjective judgment, the UAV's intelligent emotional mode should be in a fear state.

Group 4: Communication between the manned and unmanned vehicles is interrupted, and the pilot's cognitive load is defaulted to be very high. The UAV has sufficient fuel, strong radar detection capability, and the terrain is suitable for UAV flight. Based on subjective judgment, the UAV's intelligent emotional mode should not be in an aggressive state.

Substituting the data from Table 4.5 into the model yields the simulation results of the UAV intelligent emotional mode, as shown in Table 4.6.

The FCM model simulation results are visually analyzed, as shown in Figure 4.10. For Group 4, under communication interruption, the UAV's

TABLE 4.6 Simulation Results of UAV Intelligent Emotional Mode

Operating Mode	**Group 1**	**Group 2**	**Group 3**	**Group 4**
Node C_9 state value	0.829	0.4782	0.3237	0.5506
Intelligent emotional mode	Aggressive	Relaxed	Fear	Relaxed

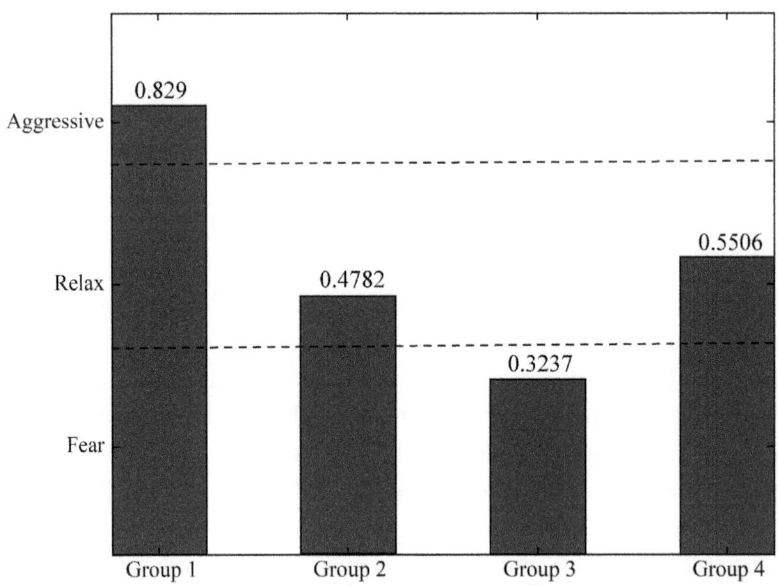

FIGURE 4.10 Visualization of FCM model simulation results for UAV intelligent emotional mode

intelligent emotional mode is in a relaxed state, consistent with the subjective judgment before simulation reasoning. The other three groups of experiments are also consistent with subjective judgments, preliminarily verifying the effectiveness and usability of the model.

4.4 CHAPTER SUMMARY

This chapter addresses the requirements for future cooperative missions between advanced manned aircraft and high-end UAVs by proposing a cognitive and interactive cooperative decision-making mechanism. The main work accomplished includes the following:

(1) Established a framework for the cognitive and interactive cooperative decision-making mechanism between manned and unmanned aerial vehicles.

(2) Introduced the classification principle based on SVM after analyzing the advantages and disadvantages of various evaluation methods.

(3) Defined the concept of the UAV intelligent emotional mode and conducted modeling and case simulation analysis.

CHAPTER 5

Manned Aircraft Pilots' Workload and Cognitive Load Evaluation Models

In this chapter, using the flight task and N-back task as a backbone, physiological indicators such as eye movement, brain signals, and heart rates are selected to establish an online SVM-based evaluation model for manned aircraft pilots' workload and cognitive load identification, while conducting further analysis and optimization.

5.1 MANNED AIRCRAFT PILOTS' WORKLOAD EVALUATION MODEL

5.1.1 Workload Experiments Based on Flight Tasks

The experiments of high or low workload were set up according to different flight tasks. During the experiments, subjects were required to complete flight tasks following predetermined routes and different maneuver requirements on the map. In the low-workload experiment, subjects flew along a zigzag route. In the high-workload experiment, subjects followed an "eight" flight path while performing coordinated left and right roll maneuvers. Experimental details are shown in Figure 5.1.

(a) Low-workload flight route

(b) High-workload flight route

DOI: 10.1201/9781003791713-5

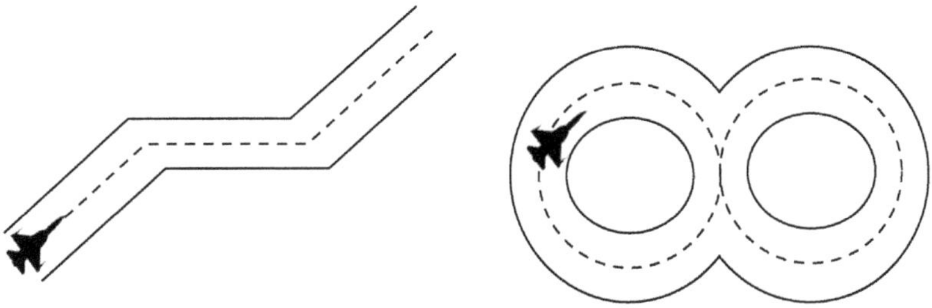

FIGURE 5.1 Schematic diagram of different flight tasks

FIGURE 5.2 Physiological sensors (recorder)

The entire flight task duration was three minutes, with the high- and low-workload flight tasks conducted alternately. During the flight mission, the subject's eye movements and multiparameter physiological data were recorded in real time. After the continuous flight time reached 20 minutes, a 10-minute rest was taken to avoid the impact of fatigue and lack of concentration on the experiment.

5.1.2 Experimental Instruments and Equipment

The physiological data acquisition instruments in the experiments mainly included physiological sensors and eye trackers, which are shown in Figures 5.2 and 5.3, respectively.

The physiological recorder is an integrated hardware and software system for physiological data acquisition, which consists of multiple physiological measurement sensors and mainly includes electromyography

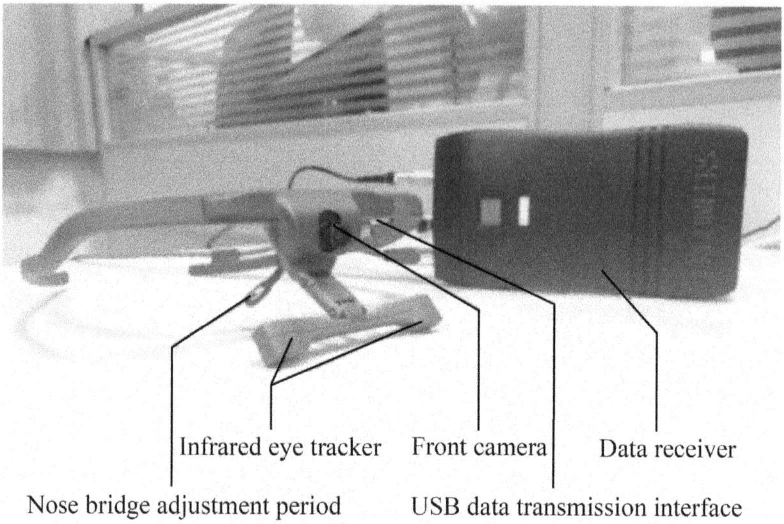

FIGURE 5.3 Eye trackers

(EMG) sensors for measuring muscle conductivity, respiratory belt sensors for measuring abdominal and thoracic respiration rates, temperature sensors for measuring body temperature, and infrared blood oxygen sensors for measuring heart rate, electrocardiogram (ECG), and blood oxygen saturation. These types of data, which are collected by the physiological recorders, are described in Table 5.1.

The eye trackers primarily use an infrared camera to record the subject's raw eye movements and status data and calculate corresponding eye movement metrics. The data types collected by the eye trackers are shown in Table 5.2.

5.1.3 Eye Movement and Physiological Indicator Selection and Acquisition

Due to limitations in the experimental design and application scope, not all the aforementioned eye movement and physiological indicators can be used for cognitive workload assessment and modeling. A rigorous selection step is necessary.

5.1.3.1 Eye Movement Indicators Selection

The eye movement indicators primarily used for cognitive workload assessment in the experiment are listed in Table 5.3.

TABLE 5.1 Different Data Types Collected by Various Sensors

Data Name	Data Type and Corresponding Explanation
EMG (mV)	Electromyography signal, that is, the spatiotemporal superposition of Motor Unit Action Potentials (MUAPs) in numerous muscle fibers. Surface EMG (SEMG) is the combined effect of superficial muscle EMG and electrical activity on nerve trunks at the skin surface, reflecting neuromuscular activity to some extent.
RIP (che) (mV)	Respiratory inductance plethysmography (chest), monitoring respiratory frequency through voltage changes in the respiratory belt sensor.
RIP (abd) (mV)	Respiratory inductance plethysmography (abdomen), monitoring respiratory frequency through voltage changes in the respiratory belt sensor.
HR (n/min)	Heart rate (i.e., the number of heart beats per minute).
SaO_2(%)	Blood oxygen saturation, the percentage of oxygenated hemoglobin (HbO_2) capacity bound to oxygen in the blood relative to the total capacity of bindable hemoglobin (Hb), i.e., blood oxygen concentration.
ECG (mV)	Electrocardiogram, that is, the comprehensive reflection of the electrical activity of countless myocardial cells in the heart.
Temperature (°C)	Temperature, measuring the subject's real-time body surface temperature.

TABLE 5.2 Data Types Collected by the Eye Tracker

Data Details	Detailed Explanations
Duration (s)	Total experiment time
Number of glances	Number of glances toward a specific area of interest (AOI)
Number of glances (>2s)	Number of glances toward a specific AOI lasting more than 2 seconds
Total glance time (s)	Total glance time during the experiment
Mean glance duration (s)	Average glance duration in the experiment
Glance rate (1/s)	Glance rate per second
AOI attention ratio	Percentage of glance time
Maximum glance duration (s)	Maximum glance duration per second
Minimum glance duration (s)	Minimum glance duration per second
Glance location probability (%)	Percentage of glance points
Link value probability (%)	Association ratio of glances between two different AOIs
Percentage transition time (s)	Percentage of AOI transition time
Horizontal eye activity (pixel)	Horizontal eye movement activity level

(*Continued*)

TABLE 5.2 *(Continued)* Data Types Collected by the Eye Tracker

Data Details	Detailed Explanations
Vertical eye activity (pixel)	Vertical eye movement activity level
PERCLOS left (%)	Eyelid closure degree (left eye)
PERCLOS average (%)	Average eyelid closure degree
PERCLOS right (%)	Eyelid closure degree (right eye)
Mean fixation duration left/right (ms)	Average fixation duration (left/right)
Number of fixations left/right	Number of fixations (left/right)
Mean saccade duration left/right (ms)	Average saccade duration (left/right)
Mean saccade angle left/right (deg)	Average saccade angle (left/right)
Number of saccades left/right	Number of saccades (left/right)
Time to first glance (s)	Time to first glance

TABLE 5.3 Eye Movement Indicators Used for Workload Assessment

Eye Movement Indicator Category	Specific Indicators
Fixation eye movement indicators	Fixation duration (ms)
	Number of fixations
Saccade eye movement indicators	Saccade duration (ms)
	Number of saccades
	Saccade angle (deg)
Other eye movement indicators	Pupil area (pixel)
	Blink rate (%)

5.1.3.2 Physiological Indicators Selection

Among the seven multiparameter physiological indicators in Table 5.1, six were selected, excluding temperature. The reasons for excluding the temperature indicator are (1) temperature lacks the sensitivity and specificity of other physiological indicators for assessing workload and (2) the experiment was conducted indoors with a fixed measurement time and environment; thus, the subject's body temperature would not change significantly.

5.1.3.3 Eye Movement and Physiological Indicators Acquisition

Based on the earlier workload experiment design, eye movement and multiparameter physiological data were collected. Each subject's experiment lasted for 20 minutes, followed by a 10-minute rest before the next session, as shown in Figure 5.4. After the entire experiment, a total of 80 datasets were collected (40 low workload and 40 high workload).

FIGURE 5.4 Eye movement and physiological indicators acquisition process

5.1.4 Eye Movement and Physiological Data Processing

After collecting eye movement and physiological data, the data need to be processed. First, outlier removal and data normalization were conducted. Next is feature selection. Considering the limited variety of eye movement and physiological data, selecting indicators with high sensitivity is sufficient during the feature selection stage. Finally, based on the selected features, a classification algorithm is used for cognitive computation and classification, followed by result validation. The specific process is shown in Figure 5.5.

5.1.4.1 Eye Movement and Physiological Data Preprocessing

The data preprocessing stage mainly includes outlier removal and normalization. Outlier processing uses the 3σ-method, removing values where the absolute difference from the mean after data standardization is greater than three times the standard deviation. Normalization processing maps each feature set $X=\{x_1,x_2,\cdots,x_n\}$ to [0,1] to eliminate the impact of different units on dimensionality reduction and classification results. The method is as follows:

$$x_i=\frac{x_i-x_{\min}}{x_{\max}-x_{\min}} \tag{5-1}$$

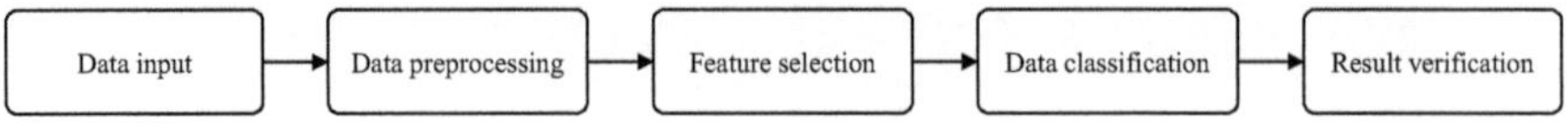

FIGURE 5.5 Specific process of eye movement and physiological data processing

TABLE 5.4 Comparison of Applicable Conditions for Several T-Test Methods

Analysis Method	Applicable Conditions	Example Illustration	Data Types
One-sample T-test	Test whether the sample mean differs from a known population mean.	Whether a class's exam scores (X) match the grade's scores (Y)	X: Quantitative Y: Quantitative
Independent samples T-test	Tests whether the means of two independent samples differ significantly from each other.	Whether exam scores (X, Y) from two different classes are consistent	X: Quantitative Y: Quantitative
Paired samples T-test	Tests whether the means of two related samples differ significantly from each other.	Whether a class's two exam scores (X, Y) are consistent	X: Quantitative Y: Quantitative

5.1.4.2 Eye Movement and Physiological Data Feature Selection

To improve classification performance, it is necessary to reduce the impact of insensitive indicators before classification. First, sensitivity analysis is performed on the data by calculating the sensitivity of each eye movement and multiparameter physiological indicator under high- and low-workload conditions and selecting indicators with high sensitivity for classification. The benefits of sensitivity analysis for indicators are twofold. First, it can reduce feature dimensionality, minimizing the impact of high dimensionality on classification performance. Second, the filtered features have good sensitivity, which can improve classification accuracy. Since classification is a binary problem (high and low workload), data sensitivity analysis is performed based on the T-test.

The T-test is used to study whether there is a significant difference between two sets of data by comparing their means. Different T-test methods correspond to different usage conditions, mainly including the one-sample T-test, independent samples T-test, and paired samples T-test, as shown in Table 5.4.

The sample data from the experiment are eye movements and multiparameter physiological data from the same subject under high- and low-workload conditions, representing a comparison of the same subject under different experiments. Therefore, the paired samples T-test should be used. Using statistical methods usually requires the original data to meet three conditions, including independence, normality, and homogeneity of variances. Since the calculation process of the paired samples T-test involves subtracting the paired data to obtain differences and comparing this set of differences to 0, if the significance level $P < 0.05$, the difference is considered not equal to 0, meaning the two sets of data from the before and after measurements are not equal. The test here compares the difference to 0, so there is no normality requirement for the original data, nor is homogeneity of variances required. The only requirement for the paired samples T-test is that the sample differences follow a normal distribution.

The principle of the paired samples T-test is as follows:

(1) First, state the null hypothesis and alternative hypothesis:

$$H_0 : \mu_d = 0, \quad H_A : \mu_d \neq 0 \tag{5-2}$$

where μ_d is the population mean of the paired data differences d, equal to the difference between the population means μ_1 and μ_2 of the two samples:

$$\mu_d = \mu_1 - \mu_2 \tag{5-3}$$

The stated null and alternative hypotheses are equivalent to

$$H_0 : \mu_1 = \mu_2, \quad H_A : \mu_1 \neq \mu_2 \tag{5-4}$$

(2) Calculate the t value using the formula:

$$t = \frac{\bar{d}}{S_{\bar{d}}}, \quad df = n - 1 \tag{5-5}$$

where $S_{\bar{d}}$ is the standard error of the difference, calculated as follows:

$$S_{\bar{d}} = \frac{S_d}{\sqrt{n}} = \sqrt{\frac{\sum (d - \bar{d})^2}{n(n-1)}} = \sqrt{\frac{\sum d^2 - \left(\sum d\right)^2}{n(n-1)}} \tag{5-6}$$

where d is the difference between the two sample data points:

$$d_j = x_{1j} - x_{2j}, \quad (j = 1, 2, \cdots, n) \tag{5-7}$$

$$\bar{d} = \sum d_j / n \tag{5-8}$$

where S_d is the standard deviation of d, and n is the number of experimental repetitions.

(3) Look up the critical t value and make a statistical inference.

Look up the critical t value $t_{0.01}(n-1)$ based on $df = n-1$, compare the absolute value of the calculated t value with it, and decide to accept or reject the null hypothesis. First, a normality test was performed on the differences between the high- and low-workload data. The test results showed that each difference followed a normal distribution. The sample data met the prerequisite requirements for the paired samples T-test. The T-test results are shown in Table 5.5.

Table 5.5 shows that the significance results for blink rate, pupil area, EMG, respiration rate (chest), and respiration rate (abdomen) are less than 0.01 under high-/low-workload conditions. Therefore, the null

TABLE 5.5 Paired Samples T-Test Results for Eye Movement and Physiological Indicators

	Paired Differences			
Paired Sample	**Mean**	**Standard Deviation**	***t*-value**	**Significance Result**
Blink rate	−5.50	4.31	−4.60	0.001*
Fixation duration	96.85	149.03	2.34	0.037
Number of fixations	45.67	74.49	2.21	0.047
Saccade duration	71.26	111.75	2.29	0.040
Saccade angle	58.46	93.11	2.26	0.043
Number of saccades	64.86	102.43	2.28	0.041
Pupil area	−89.02	85.16	−3.76	0.003*
EMG	4.06	2.65	9.679	0.000*
Respiration rate (chest)	−5.05	1.73	−18.369	0.000*
Respiration rate (abdomen)	−5.05	1.73	−18.369	0.000*
Heart rate	−0.04	0.51	−0.494	0.624
Blood oxygen saturation	0.10	0.81	0.781	0.440

* $P<0.01$.

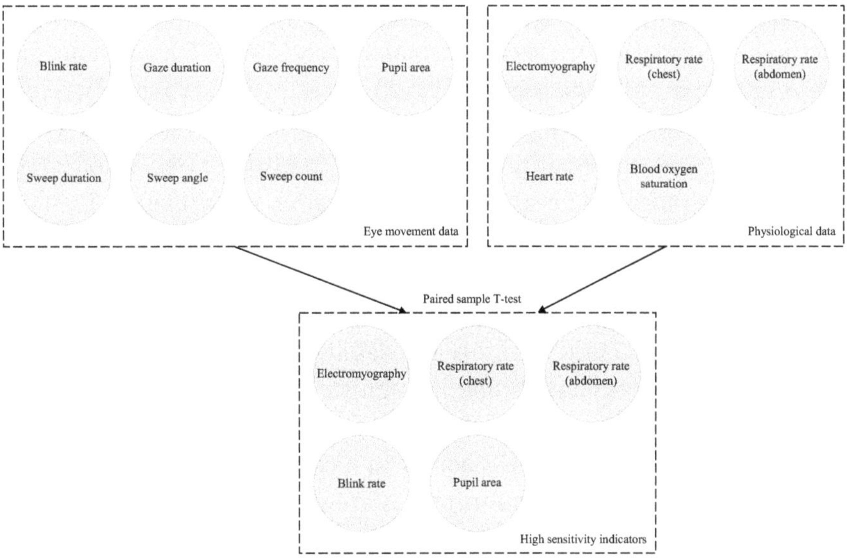

FIGURE 5.6 Indicator screening process

hypothesis is rejected, and the alternative hypothesis is accepted. That is, these five indicators show significant differences under high-/low-workload conditions and can be considered highly sensitive indicators suitable for workload assessment. Thus, these five indicators are used for data classification and validation. The screening process is shown in Figure 5.6.

5.1.4.3 Eye Movement and Physiological Data Classification and Validation

After screening for highly sensitive indicators, the SVM model is used to classify these selected high-sensitivity indicators. Considering the limited sample size (80 sets), *K*-fold cross-validation ($K = 4$) is used to comprehensively evaluate the model. The advantages of *K*-fold cross-validation (CV) are as follows: (1) Compared to the traditional holdout method, *K*-fold CV can fully cover the entire dataset, and each sampling is non-repetitive, meaning each instance has only one chance to be assigned to the training or test set in each iteration. (2) *K*-fold CV can generate multiple training and test set combinations, making it suitable for evaluating classification performance on datasets with limited sample sizes. *K*-fold

CV (especially with five or ten folds) is the most commonly applied split method. Considering the experimental sample size of 80 sets, four-fold CV is used. The entire dataset is divided into four equal subsets, each containing 20 samples. During each iteration, three subsets are used as the training set, and the remaining one subset is used as the test set to test the classification model and evaluate the classification performance. The final result is the average performance over all subsets. The principle of *K*-fold cross-validation is shown in Figure 5.7.

After applying CV, the results need to be represented appropriately. Relying solely on accuracy to judge the classification results of imbalanced data is insufficient, as it does not distinguish the misclassification cost of minority and majority class samples. Typically, a confusion matrix can be used to analyze the model's ability to identify different classes. For a binary classification problem, instances are divided into a positive class and a negative class. When using a classifier for classification, four situations occur: If an instance is positive and predicted as positive, it is called a true positive (TP). If an instance is positive but predicted as negative, it is called a false negative (FN). If an instance is negative but predicted as positive, it is called a false positive (FP). If an instance is negative and predicted as negative, it is called a true negative (TN). A matrix expressing all indicator parameters in the binary classification problem is the confusion matrix, as shown in Figure 5.8.

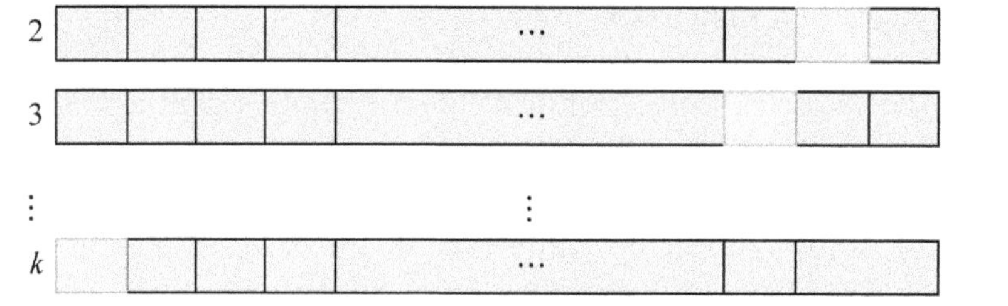

FIGURE 5.7 Schematic diagram of *K*-fold cross-validation principle

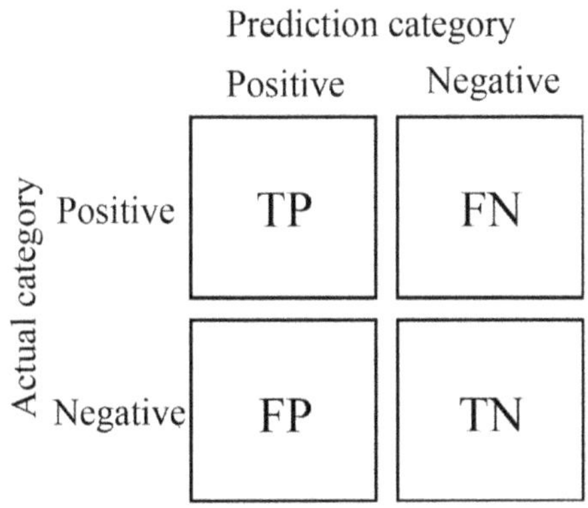

FIGURE 5.8 Confusion matrix

Furthermore, some evaluation metrics can assess the classification performance of the model, including error rate (ERR), accuracy (ACC), precision (PRE), recall (REC), F-score, receiver operating characteristic (ROC) curve, and area under the curve (AUC). Their specific definitions and explanations are as follows:

(1) Error Rate: The ratio of incorrectly predicted samples to the total number of samples.

$$\mathrm{ERR} = \frac{\mathrm{FP} + \mathrm{FN}}{\mathrm{FP} + \mathrm{FN} + \mathrm{TP} + \mathrm{TN}} \tag{5-9}$$

(2) Accuracy: The ratio of correctly predicted samples to the total number of samples.

$$\mathrm{ACC} = \frac{\mathrm{TP} + \mathrm{TN}}{\mathrm{FP} + \mathrm{FN} + \mathrm{TP} + \mathrm{TN}} = 1 - \mathrm{ERR} \tag{5-10}$$

(3) Precision: The ratio of true positive samples to the samples predicted as positive.

$$\mathrm{PRE} = \frac{\mathrm{TP}}{\mathrm{TP} + \mathrm{FP}} \tag{5-11}$$

(4) Recall: Recall is also known as "sensitivity", representing the proportion of positive samples that are correctly predicted. Precision and recall are often used together to reflect the classification performance of a model.

(5) F-score: Combines precision and recall into a single evaluation metric. The F-score is calculated based on the harmonic mean of precision and recall, giving more weight to lower values. Thus, a classifier only achieves a high F-score when both precision and recall are high.

$$F = \frac{2}{\frac{1}{\text{PRE}} + \frac{1}{\text{REC}}} = \frac{2(\text{PRE} \cdot \text{REC})}{\text{REC} + \text{PRE}} \tag{5–12}$$

(6) ROC curve: A metric for selecting classification models based on performance indicators like true positive rate (TPR) and false positive rate (FPR). TPR and FPR are two useful indicators for classification problems with class imbalance, reflecting the proportion of true positives among actual positives and false positives among actual negatives.

$$\text{TPR} = \frac{\text{TP}}{\text{FN} + \text{TP}} \tag{5–13}$$

$$\text{FPR} = \frac{\text{FP}}{\text{FP} + \text{TN}} \tag{5–14}$$

Generally, there is a diagonal line in the middle of the ROC curve, representing the performance of a completely random classifier. The target classifier's curve should be as far from it as possible.

(7) AUC value: Refers to the area under the ROC curve. The area value usually ranges between 0.5 and 1. The reason for using the AUC value as an evaluation criterion is that the ROC curve sometimes cannot clearly indicate which classifier performs better. As a numerical value, a classifier with a larger corresponding AUC value performs better. The ideal AUC value is 1, and the AUC value for random selection is 0.5. The confusion matrices for each subset are shown in Figure 5.9.

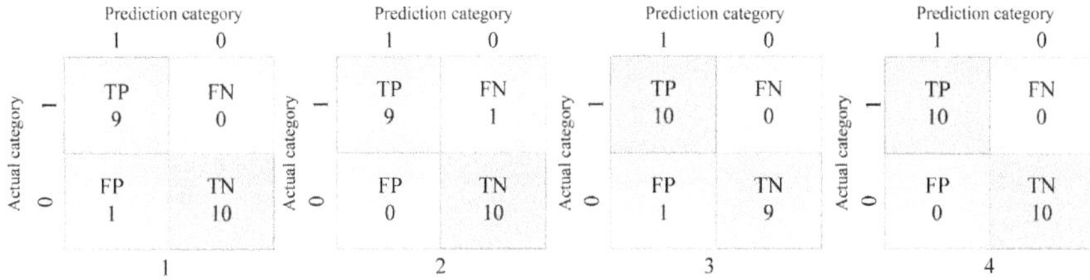

FIGURE 5.9 Confusion matrices for each subset in workload assessment

TABLE 5.6 Classification Performance for Each Subset in Workload Assessment

	Subset				
Evaluation Metric	**1**	**2**	**3**	**4**	**Average**
Error rate	0.05	0.05	0.05	0	0.0375
Accuracy	0.95	0.95	0.95	1	0.9625
Precision	0.9	1	0.9	1	0.95
Recall	1	0.9	1	1	0.975
F-score	0.947	0.947	0.952	1	0.9615
AUC Value	0.98	0.96	0.95	1	0.9725

After identifying the confusion matrices, various classification evaluation metrics are calculated, as detailed in Table 5.6. The ROC curves for each subset are shown in Figure 5.10.

Based on the earlier classification evaluation metrics, a systematic evaluation of the SVM's classification performance can be conducted. From the comprehensive evaluation metrics of the F-score and AUC value, the model shows promising classification performance for the screened eye movements and multiparameter physiological data under high-/low-workload conditions, with all positive indicators above 0.94.

After validating the effectiveness and classification accuracy of the SVM model on the experimental data, all sample data are used as training samples to obtain the final hyperplane parameters, the normal vector w, and the intercept b, for the final SVM classification assessment model. In practical application, the function $f(x) = w \cdot x + b$ can be applied to determine the classified category based on its output relationship with 0. If the output is less than 0, then it is categorized as high workload; otherwise, if it is greater than 0, it is considered low workload.

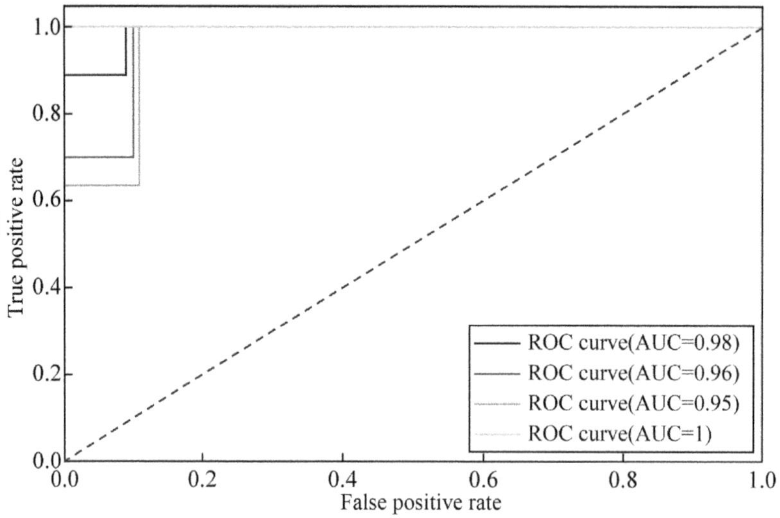

FIGURE 5.10 ROC curves for each subset in workload assessment

5.2 MANNED AIRCRAFT PILOTS' COGNITIVE LOAD ASSESSMENT MODEL

5.2.1 N-Back Task

In recent years, with the rise of cognitive neuroscience, the *N*-back task has been developed and has gradually become a relatively stable experimental paradigm, widely used to study the cognitive and neural mechanisms of working memory.

In the N-back task, a series of stimuli are presented sequentially. The subject's task is to make a "match" judgment on a target feature (such as matching letters, spatial orientations, shapes, or colors) for each presented stimulus; that is, to judge whether the currently presented item matches the item presented N trials earlier (i.e., the target item). According to previous theories, the N-back task involves at least three basic processes:

(1) Encoding each item in the presented sequence

(2) Maintaining the target stimulus and task-related information in the memory system

(3) Matching the current stimulus with the target stimulus

In the 1-back task, the target stimulus is the item preceding the current item. In the 2-back task, the target stimulus is the item presented two trials earlier. Currently, the most commonly used tasks in the literature are the 1-back and 2-back tasks because the 0-back task is simply a repetition of the current item, and tasks beyond 2-back usually impose excessive cognitive load, preventing the cognitive load experiment from proceeding normally. Therefore, the 1-back and 2-back tasks are used here as low- and high-cognitive-load experiments. An example is shown in Table 5.7 and Figure 5.11.

5.2.2 Cognitive Load Assessment Experiment Based on N-Back Task

The cognitive load assessment experiment is based on 1-back and 2-back tasks, used for low- and high-cognitive-load experiments. The specific experimental design is that the entire experiment lasts 6 minutes and 30 seconds, consisting of three repeated experimental sessions. Each experimental session lasts 2 minutes and 10 seconds. The first 10 seconds are the preparation phase, during which subjects adjust their personal state, and the subsequent 2 minutes are the *N*-back task. In the task, 60 random numbers ranging from 1 to 9 appear sequentially within 2 minutes, with an interval of 2 seconds between each number. Among the 60 random

TABLE 5.7 N-Back Task Experiment Example Table

Number Sequence	6	9	1	7	0	8	4	3	5	2
0-back task	6	9	1	7	0	8	4	3	5	2
1-back task	-	6	9	1	7	0	8	4	3	5
2-back task	-	-	6	9	1	7	0	8	4	3

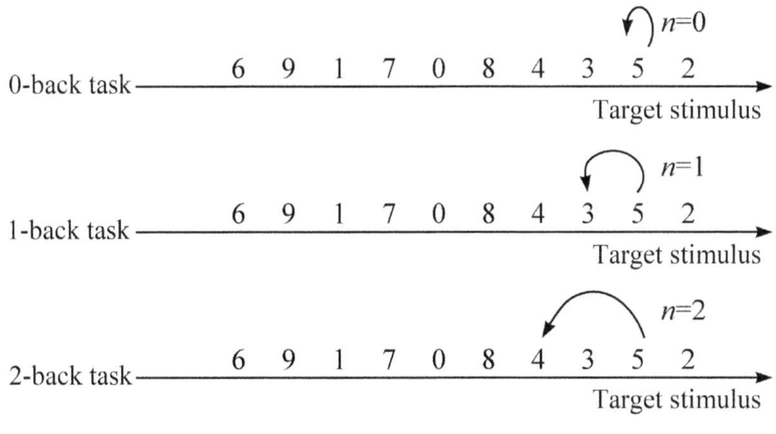

FIGURE 5.11 N-back task experimental diagram

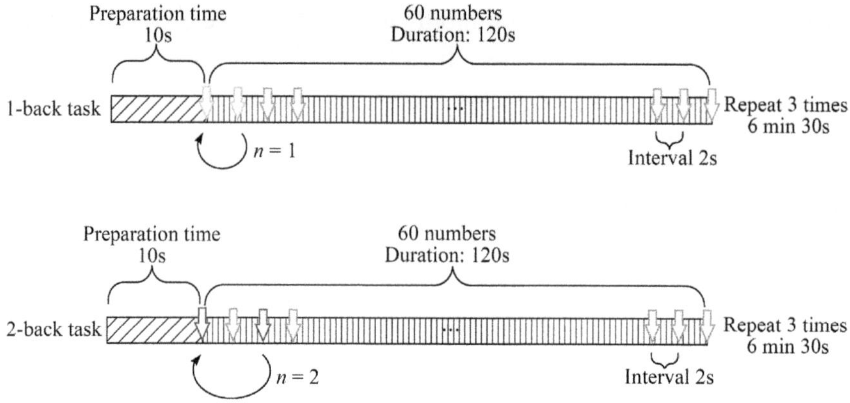

FIGURE 5.12 Cognitive load experimental procedure

numbers, 10 conform to the *N*-back task. Subjects need to judge whether the current number matches the number presented *n* trials earlier. After each group experimental session, the subject's performance accuracy rate is calculated. The experimental procedure is shown in Figure 5.12.

5.2.3 Experimental Software and Instrumentation Based on the N-Back Task

The software adopted was the *N*-back application developed by the University of Bristol, which included tasks with letters, numbers, patterns, and allowed customization. In the experiment, the *N*-back task was customized to $n = 1$ and 2, with 60 random numbers and an interval of 2 seconds between numbers. Each trial consists of 10 numbers, with a preparation time of 10 seconds.

The NE wireless EEG device used in the experiment supports up to 20 channels of sensors, with optional sampling frequencies of 500 Hz to 1000 Hz. EEG acquisition supports both dry and wet electrodes, as shown in Figure 5.13. The data types collected by the device are shown in Table 5.8.

5.2.4 Eye Movement and EEG Indicator Selection and Acquisition

5.2.4.1 Eye Movement Indicators Selection

The eye movement indicators used for cognitive load assessment are shown in Table 5.3.

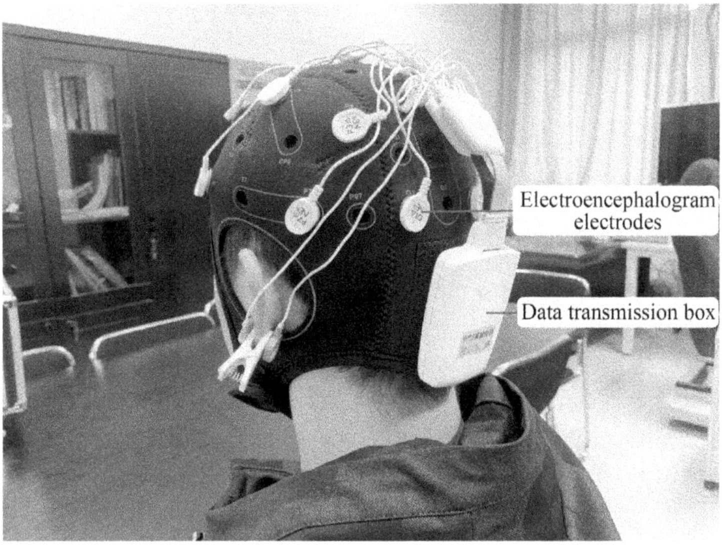

FIGURE 5.13 NE wireless EEG device

TABLE 5.8 Data Types Collected by the NE Wireless EEG Device

Data Name	**Data Type Introduction/Explanation**
Raw data all channels (μV^2)	Raw data from all channels
Filtered [2~40] Hz data (μV^2)	Filtered data (2–40 Hz)
Spectral response analysis (FFT) [μV^2/sprt(Hz)]	Spectral analysis via fast Fourier transform (FFT)
Spectral response analysis (PSD) (dB/Hz)	Spectral analysis based on power spectral density (PSD)
Band power analysis (μV^2)	Frequency band power analysis

5.2.4.2 EEG Indicator Selection and Electrode Setup

Among EEG indicators, the frequency band power of the EEG signal is usually taken as the main analytical indicator. The EEG frequency band power from 2–40 Hz is divided into five frequency bands, including Delta, Theta, Alpha, Beta, and Gamma. Internationally, Greek letters δ, θ, α, β, γ are used uniformly. These five frequency bands correspond to specific brain activity functions and different cognitive characteristics during cognitive processing. The division of EEG frequency band power is shown in Figure 5.14.

FIGURE 5.14 Division of EEG frequency band power

Detailed interpretations of each frequency band are as follows:

(1) Delta band (δ): Amplitude of 20–200 μV with a frequency of 2–4 Hz. This band often has the highest amplitude and slowest fluctuations and typically can be found in adults during slow-wave sleep. δ waves are often used to assess sleep depth; that is, the stronger the rhythm is, the deeper the subject sleeps.

(2) Theta band β: Amplitude range of 10–100 μV with a frequency of 5–7 Hz. This band is typically seen in young children and is also often detected in deep meditation states in adults. θ waves become more prominent whenever a person encounters a difficult mental task.

(3) Alpha band (α): Amplitude of 10–100 μV and a frequency of 8–13 Hz. This band is a marker of relaxed brain activity. α waves generally connect the two brain hemispheres in a synchronized manner. They are most prominent when the subject is relaxed, typically appearing with eyes closed and attenuating upon eye opening or during mental activity.

(4) Beta band (β): Amplitude of 20–100 μV and a frequency of 14–30 Hz. EEG waves in this band are often symmetrically distributed on both sides of the brain and are most dominant in the frontal lobe, reflecting the brain's alert state. Higher β frequencies are associated with high levels of arousal activity in the brain. The presence of more β band waves often indicates a state of anxiety, high excitement, inability to relax, or stress.

(5) Gamma band (γ): Frequency of 30–40 Hz. This band belongs to the high-frequency components of EEG waves. γ waves are associated with cognitive functions and higher-level information processing. They play a significant role in activities related to learning, memory, and information processing.

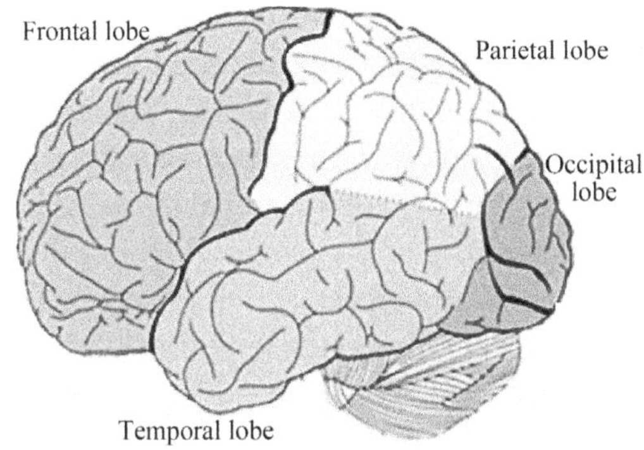

FIGURE 5.15 Diagram of brain region division

After selecting the dominant EEG indicators, optimal electrode placement must be determined based on brain region mapping to ensure effective signal acquisition, as illustrated in Figure 5.15.

The brain is generally divided into four lobes, including the frontal lobe, parietal lobe, occipital lobe, and temporal lobe, specifically:

(1) The frontal lobe, situated at the front of the brain, represents the most advanced region in human cortical development. It encompasses the motor cortex, premotor cortex, and prefrontal cortex, and its primary function is to direct goal-oriented behavior in accordance with human intention.

(2) The parietal lobe is located behind the central sulcus and before the line connecting the parieto-occipital sulcus and the preoccipital notch. Its function is related to pain, touch, taste, temperature, pressure, and logical ability. Research indicates that greater volume in the posterior parietal region correlates with enhanced logical and divergent thinking abilities. The parietal lobe's size is also associated with lifespan and with functions related to balance and coordination.

(3) The occipital lobe is a region of the cerebral cortex located at the back of the hemisphere, behind the parieto-occipital sulcus. It is the visual cortex center, and its main functions include processing visual information such as that of the primary visual cortex located in the occipital lobe.

(4) The temporal lobe is located below the lateral sulcus, above the middle cranial fossa and tentorium cerebelli, with the frontal lobe in

front, the frontoparietal lobe above, and the occipital lobe behind. Areas 41 and 42 of the superior temporal gyrus and the transverse temporal gyrus are auditory cortex areas. The posterior part of the superior temporal gyrus is the auditory speech center in the dominant hemisphere. The para-hippocampal gyrus is the olfactory/taste center. The anterior part of the temporal lobe is the psychic cortex, and human emotions and mental activities are related not only to the orbitofrontal cortex but also to the temporal lobe. The hippocampus is related to memory.

Generally, the electrode positions on the brain regions widely use the "10–20 International Standard System", which standardizes the relative mapping positions between electrodes and the naming of electrodes. This system uses even numbers to represent electrodes on the right hemisphere and odd numbers for the left hemisphere. Letters correspond to various brain regions of the cerebral cortex, as shown in Figure 5.16.

(a) Side view of the system

(b) Top view of the system

The number of electrodes for EEG acquisition often uses 16, 32, 62, 64, 128 channels, etc. The more electrodes used, the higher the requirement

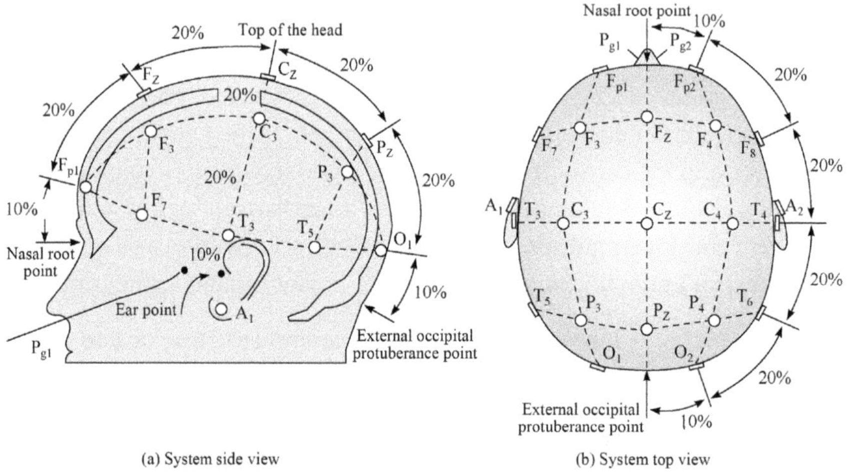

FIGURE 5.16 10–20 International standard system

TABLE 5.9 Electrodes and Their Associated Brain Regions

No.	Electrode	Brain Region	Associated Lobe	Function
1	P_7	Parietal lobe	Parietal lobe	Processes touch, pressure, temperature, pain; regulates attention; or allocates spatial attention.
2	P_4	Parietal lobe, right parietal		
4	P_z	Midline parietal		
5	P_3	Parietal lobe, left parietal		
6	P_8	Parietal lobe		
3	C_z	Central midline	Paracentral lobe	Responsible for thinking, calculation, and related to individual needs and emotions.
11	C_4	Right central		
15	C_3	Left central		
7	O_1	Left occipital	Occipital lobe (Visual cortex)	Related to the integration of visual information.
8	O_2	Right occipital		
9	T_8	Temporal lobe	Temporal lobe	Mainly processes auditory stimuli and consists of subregions handling audition, language, and certain aspects of memory.
18	T_7	Temporal lobe		
10	F_8	Right prefrontal	Frontal lobe	Related to somatic movement, vocalization, language, and higher thinking activities.
12	F_4	Right frontal		
13	F_{P2}	Right frontopolar		
14	F_z	Frontal midline	Frontal lobe	Related to physical movement, pronunciation, language, and advanced mental activities.
16	F_3	Left frontal		
17	F_{P1}	Left frontopolar		
19	F_7	Left prefrontal		
20	EXT	Left (right) earlobe	—	Ground, reference signal.

for signal quality. The 20-channel EEG electrodes used in the experiment were selected based on the distribution of the standard 64-channel EEG electrodes, as shown in Table 5.9 and Figure 5.17.

5.2.4.3 Eye Movement and EEG Data Acquisition

To avoid mental fatigue, each subject's experiment lasted for 20 minutes, followed by a 10-minute rest before the next session. The data acquisition process is shown in Figure 5.18. After the entire experiment, a total of 98 sets of eye movement and EEG data were collected (including 49 low cognitive load and 49 high cognitive load).

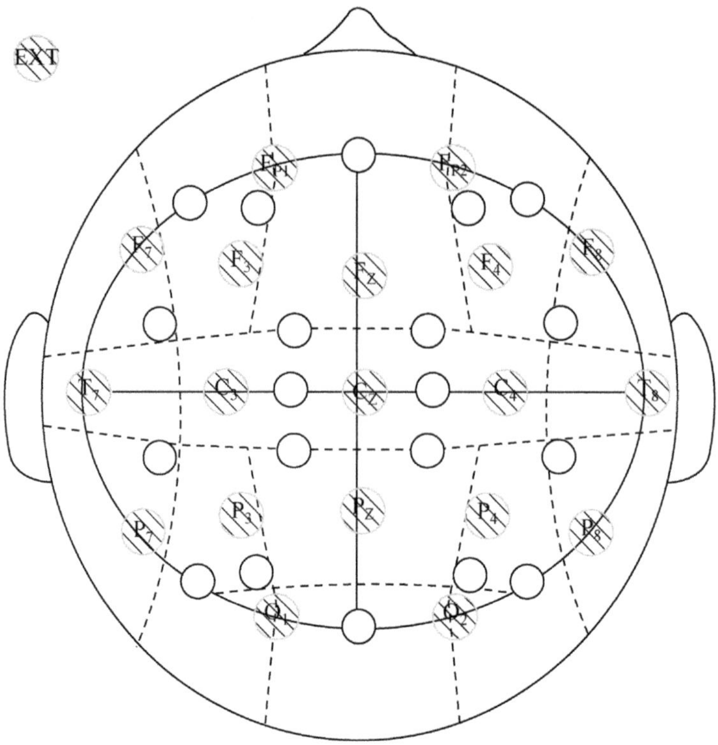

FIGURE 5.17 20-channel EEG electrode distribution

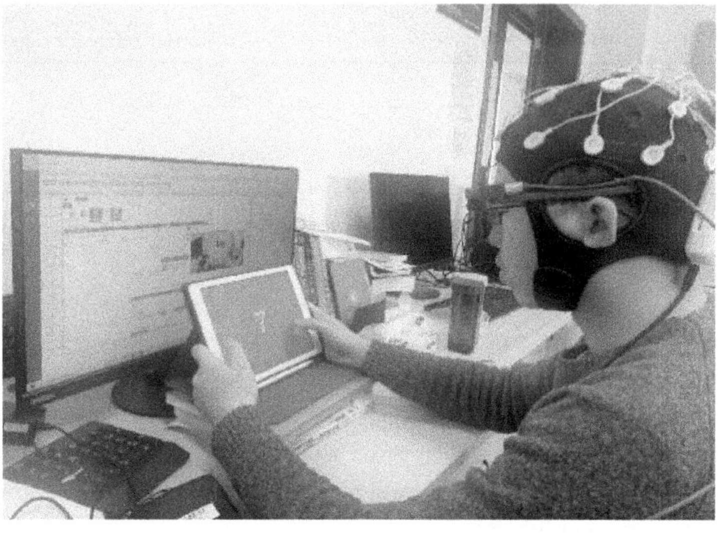

FIGURE 5.18 Data acquisition process

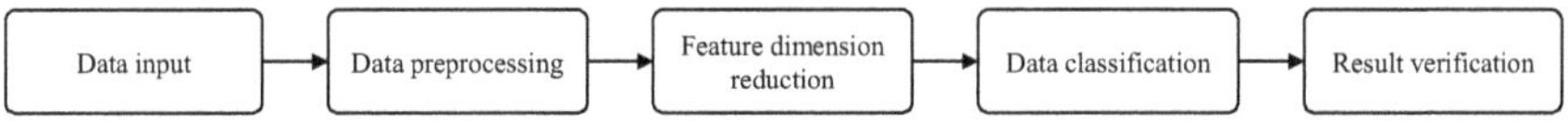

FIGURE 5.19 Eye movement and EEG data processing flow

5.2.5 Eye Movement and EEG Data Processing

After collecting eye movement and EEG signals, the data need to be processed and input into the model to obtain the cognitive load through the assessment model. The data sample consists of 98 sets, with a feature dimension of 107, including 7 selected eye movement feature indicators and 100 EEG feature indicators. Analyzing the sample size and feature dimension shows that their dimensionalities are almost the same, belonging to high-dimensional data. Direct classification would lead to problems like the "curse of dimensionality", "empty space", and "algorithm failure", especially for the experimental data sample, where the number of samples and dimensions are close, causing sample points in high-dimensional space to be extremely sparse and making computation difficult. To mitigate these effects, dimensionality reduction serves as an effective approach. The complete data processing workflow is illustrated in Figure 5.19.

5.2.5.1 Data Preprocessing

Data preprocessing includes outlier removal and normalization. For outlier removal, data for each indicator are visualized using histograms, and outliers are manually removed. Figure 5.20 shows the original data distribution for the four selected indicators, including delta wave (O_1 electrode), theta wave (F_3 electrode), blink rate, and pupil area. It is evident through data analysis that outliers exist in these indicators. After removing outliers from each indicator, the number of data samples was reduced from the original 98 sets to 90 sets.

After outlier removal, normalization is required to eliminate the influence of different scales on dimensionality reduction and classification results. This involves mapping each feature set $X = \{x_1, x_2, \cdots, x_n\}$ into [0,1], using the following formula:

$$x_i = \frac{x_i - x_{\min}}{x_{\max} - x_{\min}} \tag{5-15}$$

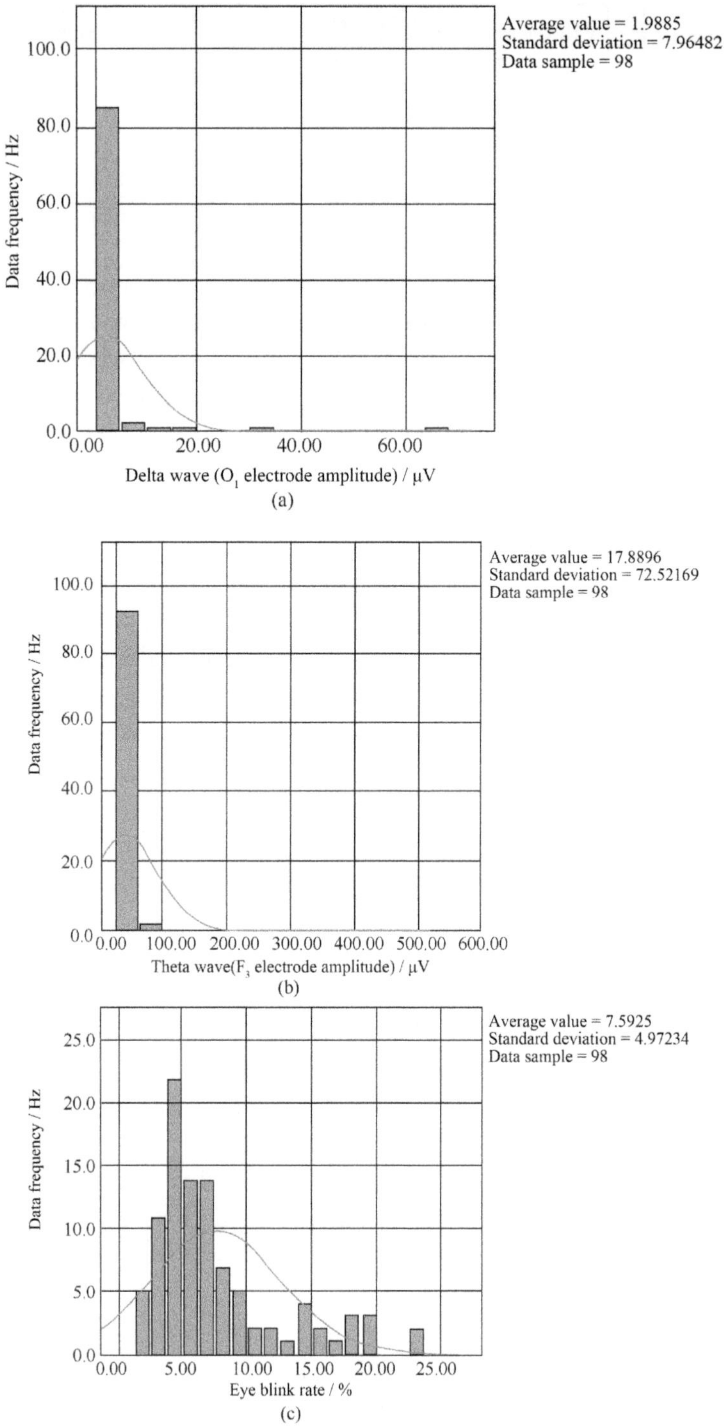

FIGURE 5.20 Original data distribution of various indicators

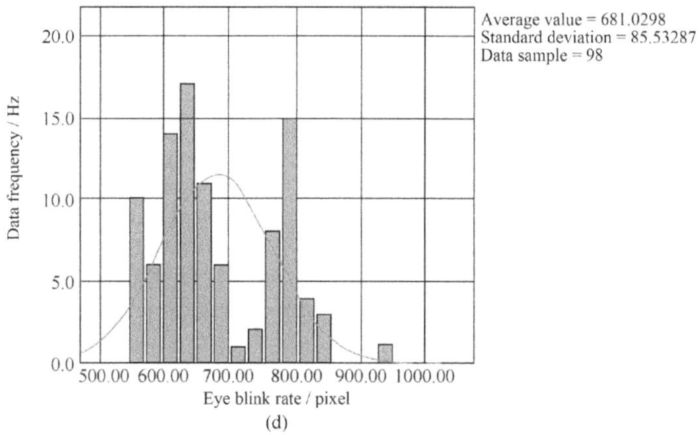

FIGURE 5.20 Continued

5.2.5.2 Data Dimensionality Reduction

Factor analysis based on principal component analysis (PCA) is used for data dimensionality reduction. Compared with PCA, factor analysis can apply rotation techniques to enhance factor interpretability, offering an advantage in terms of explanatory clarity. A detailed comparison of these two methods is shown in Table 5.10.

In summary, both PCA and factor analysis can transform existing variables into a few new variables (which contain most of the information from all original variables) for subsequent analysis. When there is a need to identify potential factors and interpret them, factor analysis is preferred. For the multidimensional EEG and eye movement data used in cognitive load assessment, the goal after dimensionality reduction is to obtain interpretable, more abstract data indicators.

The mathematical model of factor analysis is as follows: let $X_i (i = 1, 2, \cdots, p)$ contain p number of variables. If $X_i = \mu_i + a_{i1}F_1 + \cdots + a_{im}F_m + \varepsilon_i (m \le p)$, then $F_1, F_2, \cdots, F_m$ are called common factors, which are unobservable variables. $(a_{i1}, a_{i2}, \cdots, a_{im})$ is called the factor loading. ε_i is the specific factor, representing the part not included in the first m common factors, satisfying $\mathrm{Cov}(F, \varepsilon) = 0$, meaning F and ε are uncorrelated:

$$D(F) = \begin{bmatrix} 1 & & & \\ & 1 & & \\ & & \ddots & \\ & & & 1 \end{bmatrix} = I \tag{5-16}$$

TABLE 5.10 Comparison of Factor Analysis and PCA

Similarities	Differences
(1) Both methods eliminate the correlation among the original variables, simplify the structure of correlated variables, and reduce dimensionality. They construct composite indicators to rank the original data, ultimately achieving dimensionality reduction and ranking.	(1) In PCA, if the eigenvalues of the given covariance matrix or correlation matrix are unique, the principal components are generally unique, whereas in factor analysis, factors are not unique and can be rotated to obtain different factors.
(2) Both methods measure the retrieved information by variance magnitude, which is directly proportional. Weight values also correlate with information quantity, meaning weight magnitude is proportional to variance magnitude. Their objectivity lies in the fact that weights emerge algorithmically during the transformation between original and derived indicators, free from manual intervention.	(2) In PCA, the principal components are expressed as linear combinations of the data variables, whereas in factor analysis, the variables are considered as linear combinations of the factors. (3) PCA focuses on explaining the total variance of each variable, whereas factor analysis emphasizes explaining the covariance between variables.

This means $F_1, F_2, \cdots, F_m$ are mutually uncorrelated, with a variance of 1.

$$D(\varepsilon) = \begin{bmatrix} \sigma_1^2 & & & \\ & \sigma_2^2 & & \\ & & \ddots & \\ & & & \sigma_p^2 \end{bmatrix} \tag{5–17}$$

This means ε are mutually uncorrelated, their variances are not necessarily equal, and ε follows a normal distribution$\varepsilon_i\hat{}\ N(0,\sigma_i^2)$.

The previous model is represented in matrix form as follows:

$$X - \mu = AF + \varepsilon, \quad E(F) = 0 \tag{5–18}$$

$$E(\varepsilon) = 0, \quad \mathrm{Var}(F) = I \tag{5–19}$$

$$\mathrm{Cov}(F,\varepsilon) = E(F\varepsilon) = \begin{bmatrix} E(F_1\varepsilon_1) & E(F_1\varepsilon_2) & \cdots & E(F_1\varepsilon_p) \\ E(F_2\varepsilon_1) & E(F_2\varepsilon_2) & \cdots & E(F_2\varepsilon_p) \\ \vdots & \vdots & & \vdots \\ E(F_p\varepsilon_1) & E(F_p\varepsilon_2) & \cdots & E(F_p\varepsilon_p) \end{bmatrix} = 0 \tag{5–20}$$

$$\mathrm{Var}(\varepsilon) = \mathrm{diag}(\sigma_1^2, \sigma_2^2, \cdots, \sigma_p^2) \tag{5–21}$$

Factor analysis was performed on the experimental data, and the results are shown in Figure 5.21, which is also called a scree plot (i.e., a concept derived from geology). The scree plot uses eigenvalues as the vertical axis and components as the horizontal axis. The steep front part has large eigenvalues and contains more information, while the flat rear part has small eigenvalues and contains less information. It can be intuitively seen that components 1 to 5 contain most of the information; hence, the plot is steeper, while components 6 and beyond contain less information, making the plot flatter. Specific information is shown in Table 5.11.

The first column in Table 5.11 lists the components, and the experiment selected the first 12 components with eigenvalues greater than 1. The second column lists the corresponding eigenvalues, indicating the amount of variance explained. The third column lists the percentage of the total variance contained by the corresponding component. The fourth column is the cumulative percentage. In Table 5.11, components 1 to 5 have eigenvalues greater than 1 and collectively explain 81.056% of the variance. Therefore, components 1 to 5 are taken as the principal components, while the remaining components contain less information and can be ignored.

After extracting the main components, the relationship between the components and the original variables can be observed through the component matrix (Appendix Table 1). The values in the matrix are the correlation coefficients between the components and the original variables, and the larger the absolute value, the closer the relationship. From the

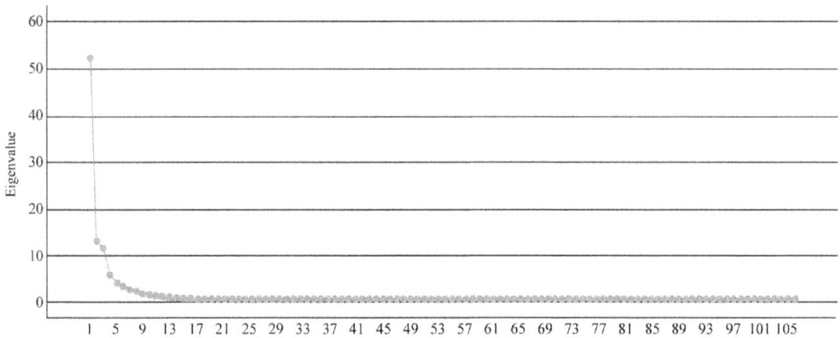

FIGURE 5.21 Factor analysis results graph

TABLE 5.11 Information Related to Components 1 to 12

Component	Eigenvalue	Variance Percentage (%)	Cumulative Percentage (%)
1	52.198	48.783	48.783
2	13.058	12.204	60.987
3	11.566	10.810	71.797
4	5.814	5.434	77.231
5	4.092	3.825	81.056
6	3.335	3.117	84.173
7	2.654	2.480	86.653
8	2.290	2.141	88.794
9	1.609	1.504	90.298
10	1.588	1.484	91.782
11	1.479	1.382	93.164
12	1.304	1.219	94.383

component matrix, it can be seen that the extracted components already have good interpretability, so factor rotation is not performed. Among the extracted components, components 1, 3, 4, and 5 have relatively clear relationships with the original variables. Taking components 1 and 3 as examples, their relationships with the original variables are represented using heatmaps, as shown in Figures 5.22 and 5.23.

By analyzing the heat maps across different frequency bands, it can be concluded that in component 1, the delta, theta, beta, and gamma frequency bands in the brain regions are active. Therefore, component 1 is closely related to delta, theta, beta, and gamma waves and can reflect a subject's alert, thinking, and vigilant state. In component 3, the alpha frequency band in the brain regions is active. Therefore, component 3 is closely related to alpha waves and can reflect a subject's relaxed state. The specific interpretations of components 1 to 5 are shown in Table 5.12.

Additionally, Appendix Table 2 shows the weight coefficient matrix for each component, which can be used to transform the original variables into the various components. After component extraction is completed, classification algorithms are used to classify the EEG and eye movement data under high and low cognitive load.

5.2.5.3 Data Classification and Result Validation

After data dimensionality reduction, an SVM classifier is used to classify the reduced-dimension eye movement and EEG data. Considering the limited sample size (90 sets), *K*-fold CV ($K = 9$) is used to comprehensively

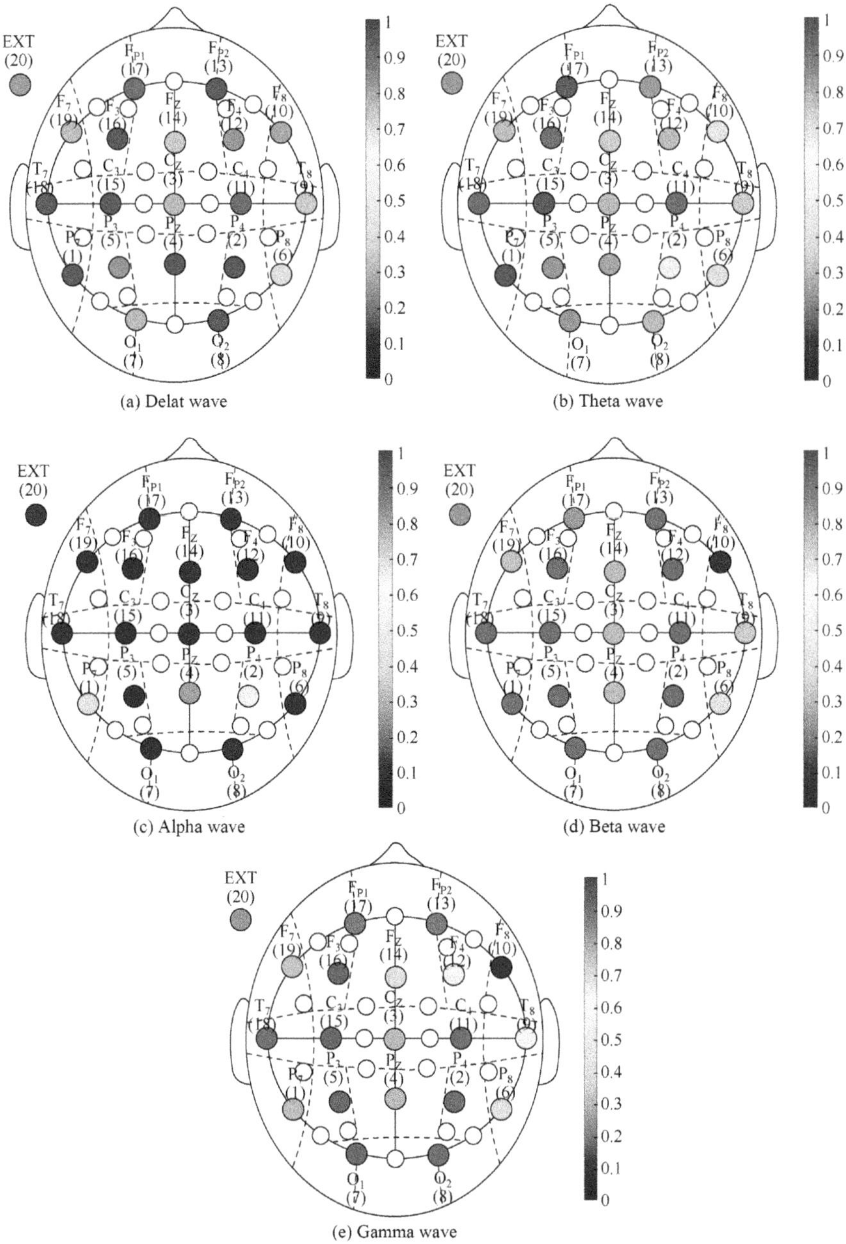

FIGURE 5.22 Relationship between component 1 and original variables

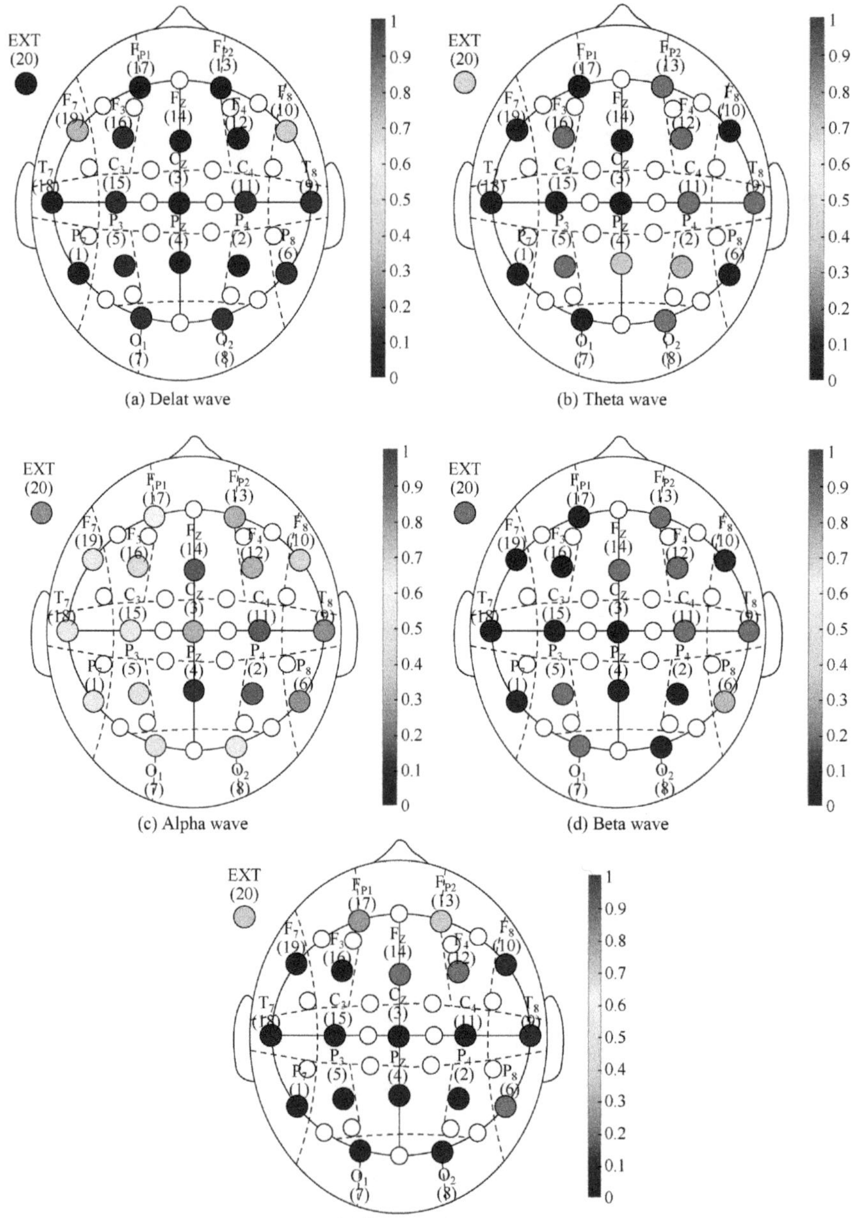

FIGURE 5.23 Relationship between component 3 and original variables

TABLE 5.12 Specific Interpretation of Components 1 to 5

Component	Related Original Variables	Interpretation of Component
1	Delta, theta, beta, gamma waves	Alert, thinking, vigilant state
2	Delta, theta, alpha, beta, gamma waves	Lacks Good Interpretability
3	Alpha waves	Relaxed state
4	Alpha waves	Relaxed state
5	Alpha waves	Relaxed state

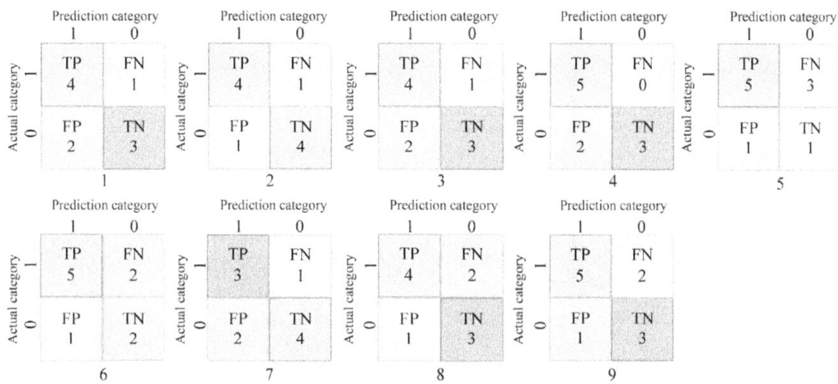

FIGURE 5.24 Confusion matrices for each subset in cognitive load assessment task

evaluate the model. The confusion matrices for each subset are shown in Figure 5.24, the classification performance is shown in Table 5.13, and the ROC curves are shown in Figure 5.25.

Based on the classification evaluation metrics in Table 5.13, a systematic evaluation of the SVM's classification performance can be conducted. From the F-score and AUC values, the model shows relatively promising classification performance for the reduced-dimension EEG and eye-movement data under high and low cognitive load conditions. To further improve classification accuracy, optimization can be performed in aspects such as feature extraction, classification model, data collection environment, personnel state, habits, and gender differences.

After validating the effectiveness and classification accuracy of the SVM model on the experimental data, all sample data are used as training samples to obtain the final hyperplane parameters, the normal vector w and the intercept b, for the final SVM classification assessment model. The

TABLE 5.13 Classification Performance for Each Subset in Cognitive Load Assessment

Evaluation Matrices	Subset 1	2	3	4	5	6	7	8	9	Average
Error rate	0.3	0.2	0.3	0.2	0.4	0.3	0.3	0.3	0.27	0.286
Accuracy	0.7	0.8	0.7	0.8	0.6	0.7	0.7	0.7	0.73	0.714
Precision	0.67	0.8	0.67	0.71	0.83	0.83	0.6	0.8	0.83	0.749
Recall	0.8	0.8	0.8	1.0	0.625	0.71	0.75	0.67	0.714	0.763
F-score	0.72	0.8	0.72	0.83	0.71	0.77	0.67	0.72	0.77	0.746
AUC Value	0.88	0.96	0.98	0.86	0.96	0.76	0.76	0.75	0.84	0.861

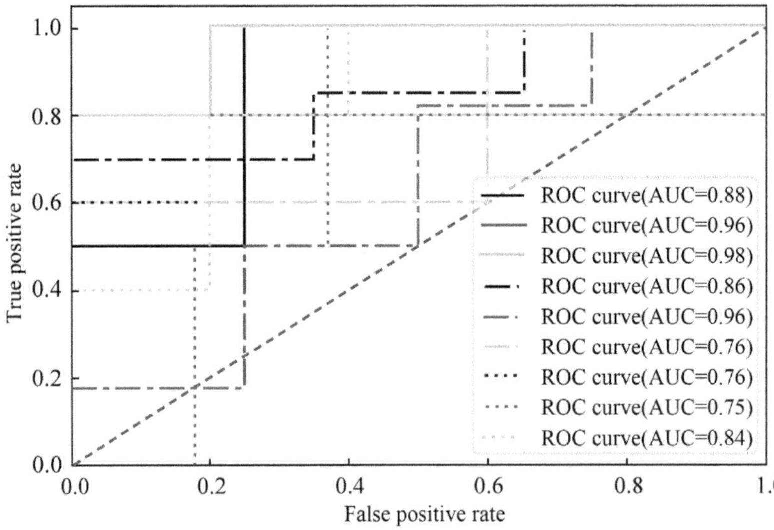

FIGURE 5.25 ROC curves for each subset in cognitive load assessment

classified category is determined based on the relationship between the decision equation $f(x) = w \cdot x + b$ and 0. If the output is less than 0, then it is categorized as high cognitive load; otherwise, if it is greater than 0, it is low cognitive load.

5.3 OPTIMIZATION OF MANNED AIRCRAFT PILOTS' WORKLOAD AND COGNITIVE LOAD ASSESSMENT MODELS

5.3.1 Model Analysis

5.3.1.1 Workload Assessment Model

The workload assessment model is based on the selected five indicators with high sensitivity under the high-/low-workload conditions, including blink rate, pupil area, EMG, respiration rate (chest), and respiration rate (abdomen). It demonstrates promising classification performance with the SVM model. From the AUC values of the *K*-fold CV (K=4), the average accuracy rate is 0.9725, the minimum result is 0.95, and the maximum result is 1.0. It has good classification performance and relatively stable results; therefore, no further optimization is performed on the workload model. The reasons are mainly the following three points: (1) The number of target categories for classification is limited, being a binary classification problem (high/low workload), making it easier to achieve good classification performance. (2) The feature dimension of the input for classification is not high (5 dimensions), also making it easier to achieve good classification performance. (3) The sample features for workload were pre-screened, and these features already have good separability under high- and low-workload conditions, allowing for good classification performance.

5.3.1.2 Cognitive Load Assessment Model Analysis

The cognitive load assessment model is based on EEG and eye movement data and is classified using the SVM model. From the AUC results of the *K*-fold CV ($K = 9$), the average result is 0.861, the minimum result is 0.75, and the maximum result is 0.98. The difference between the maximum and minimum is 0.23, indicating poor stability of the classification performance and less-than-ideal results. Therefore, the focus is on optimizing the cognitive load model.

Before optimizing the model, the reasons for the poor classification performance of the cognitive load model are first analyzed: (1) From the perspective of classification samples, the EEG data have a high feature dimension (100 dimensions). Compared to the sample size (98 sets), this can easily lead to the "curse of dimensionality" during classification, resulting in poor classification performance. (2) From the perspective of sample data preprocessing, dimensionality reduction was performed on the high-dimensional EEG and eye movement data before classification, reducing the features from 107 dimensions to 5 dimensions. While the data dimensionality was reduced, a significant amount of information was also lost. (3) During the dimensionality reduction process, the information inherent in each feature itself was not considered, and the features were simply reduced together.

Therefore, by identifying the reasons for the poor classification performance earlier, improvement methods are proposed to enhance the classification effect.

5.3.1.3 Improvement Method

To improve the classification performance, it is necessary to fully consider the information and meaning inherent in each feature during data dimensionality reduction. This requires further analysis of the relationship between EEG and eye movements and cognitive workload. Among the 107-dimensional EEG and eye movement data, they can be divided into the following six types: delta band, theta band, alpha band, beta band, gamma band in the EEG signals, and the eye movement indicators. These six types of information have different manifestations and meanings under different cognitive load conditions. It is necessary to further divide the 107-dimensional features based on the significance of each feature before performing classification processing.

Referring to the research status of cognitive load assessment based on EEG and eye movement data, studies have shown that the power of frequency bands such as theta, alpha, and beta waves in EEG signals is sensitive to changes in mental workload. When a person is in a state of high alertness and performing more difficult tasks, the main components of EEG activity tend to shift toward the beta band, which has lower amplitude and higher frequency. When a person is awake but with lower alertness, alpha wave activity increases. When a person is in a drowsy state, theta waves significantly increase [79]. It is generally believed that psychological stress, active thinking, and attention all promote the shift of EEG activity toward higher frequency bands and suppress alpha wave activity. Additionally, the power of alpha and theta waves is correlated with task difficulty and cognitive load [80–84]. Regarding eye movement data [85], pupil area gradually increases with increasing cognitive load. In simulated flight tasks, pupil area and eyelid opening show significant differences under different cognitive load states [86, 87].

From the aforementioned research status, it can be seen that not all EEG and eye movement signals have significant differences under different cognitive load states. Therefore, during data dimensionality reduction and classification, data with different levels of significance should be separated, and the weight of features with greater significance for classification should be increased. The significant differences of each feature under high/low cognitive load can be determined based on paired sample T-tests. The number of features with significance is shown in Table 5.14.

From Table 5.14, it can be seen that the number of features with significant differences under high/low cognitive load is relatively high for eye movement data and the alpha band of EEG data, at 85.7% and 60%,

TABLE 5.14 Number of Significant Features in EEG and Eye Movement Data

Category	Specific Data	Number of Significant Features (Total Number)	Percentage of Significant Features
EEG	Delta band	4 (20)	20%
	Theta band	5 (20)	25%
	Alpha band	12 (20)	60%
	Beta band	3 (20)	15%
	Gamma band	4 (20)	20%
Eye Movement	Blink rate, fixation Duration, number of fixations, Saccade duration, number of saccades, saccade angle, pupil area	6 (7)	85.7%

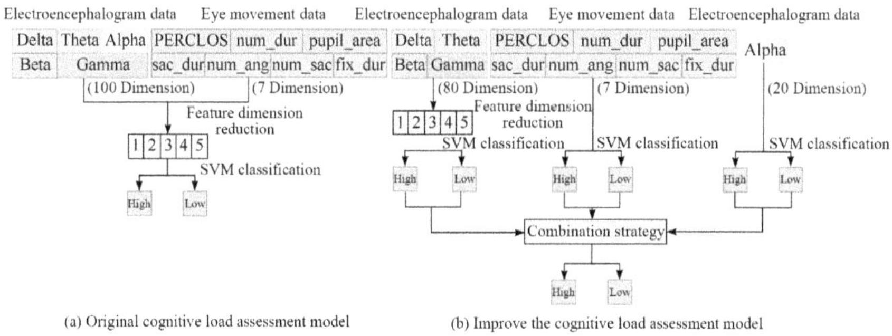

FIGURE 5.26 Original and improved cognitive load models

respectively. Therefore, the eye movement data and the alpha band of EEG data can be separated from the 107-dimensional eye movement and EEG data and classified separately using the SVM model. The remaining data can be classified using the SVM model after dimensionality reduction. Considering the situation of multiple classifiers, this section proposes using a combined classifier to improve the cognitive load model. The details are shown in Figure 5.26.

5.3.2 Combined Classifier Design

Integrating multiple methods to train a strong classifier is known as a multi-classifier ensemble or ensemble learning method [88]. The basic principle is to use several base (weak) classifiers, obtain multiple classification results, and then combine the results of these base classifiers using a certain aggregation method to get the final output. Ensemble learning methods can be used not only for classification problem integration [89] but also for regression problem integration [90], feature selection integration, and outlier detection integration, to name a few. Ensemble learning methods can be found in almost all areas of machine learning. The principle of a multi-classifier ensemble is shown in Figure 5.27. The acquired dataset is divided into training and test sets. The training set is used to train several classifiers, which are then combined through a certain strategy to form a strong learner.

There are many methods for combining the results from multiple classifiers, and they are introduced in the following. Assume T weak learners $\{h_1, h_2, \cdots, h_T\}$ are obtained. There are:

(1) Averaging method: mainly used for numerical regression prediction problems. It obtains the final prediction output by averaging the outputs of several weak learners, via

$$H(x) = \frac{1}{T}\sum_{i=1}^{T} h_i(x) \tag{5-22}$$

If a weight value w_i is assigned to each weak learner, the final prediction result is as follows:

$$H(x) = \sum_{i=1}^{T} w_i h_i(x) \tag{5-23}$$

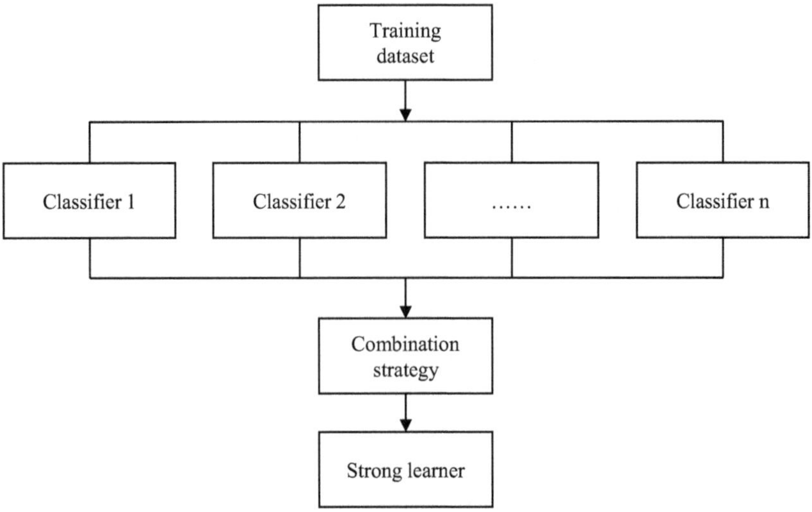

FIGURE 5.27 Multi-classifier ensemble principle diagram

Here w_i is the weight of the weak learner h_i, usually satisfying:

$$w_i \geq 0, \quad \sum_{i=1}^{T} w_i = 1 \tag{5–24}$$

(2) Voting method: typically used in the prediction phase of classification problems. It can be divided into majority voting and weighted voting. Assume the predicted classes are $\{c_1, c_2, \cdots, c_k\}$. Then for any prediction sample x, the classification results of the T classifiers are $(h_1(x), h_2(x), \cdots, h_T(x))$.

Majority voting is a relatively straightforward ensemble method. To produce a valid prediction outcome, two conditions must be met. The result must receive both the highest number of votes and more than half of the total votes cast; otherwise, the prediction is rejected.

Weighted voting is more complex than majority voting. When combining multiple classifiers through weighted voting, the classification vote from each classifier is multiplied by an assigned weight, after which the weighted votes for each class are summed. The class with the highest total is selected as the final outcome.

(3) Learning Method. Voting and averaging methods are relatively simple but may lead to significant errors during the learning stage; hence, the learning method emerged. A representative learning method is the stacking method. When using the stacking method, simply performing logical processing on the results of the weak learners is far from sufficient. A key step involves adding another layer of learners. The basic process of the stacking method is as follows. In the training phase, the outputs of the base classifiers are used as input features, and the original training labels serve as outputs to train a new classifier, called the meta-learner. In the testing phase, the base classifiers first generate predictions, which are then used as input features for the meta-learner. The meta-learner makes a final prediction based on these inputs to produce the ultimate classification result.

Among the above three methods, the voting method is the most commonly used and relatively reliable one. However, since the ensemble process of majority voting only combines the results without additional information, it cannot reflect the advantages of high-performance classifiers. In contrast, weighted voting can assign higher weights to classifiers with higher performance, thus leveraging their advantages. Therefore, the ensemble multi-classifier weighted voting method is adopted to optimize the cognitive load model.

5.3.2.1 Weighted Voting Classifier Mechanism

The weight matrix for the classifier recognition status is as follows:

$$W = \begin{bmatrix} w_{11} & w_{12} & \cdots & w_{1n} \\ w_{21} & w_{22} & \cdots & w_{2n} \\ \vdots & \vdots & & \vdots \\ w_{m1} & w_{m2} & \cdots & w_{mn} \end{bmatrix} \tag{5–25}$$

Here, m is the number of classifiers, n is the number of classes, and w_{ij} is the weight set for the i-th classifier when its output is the j-th class. The value of each element in the weight matrix is as follows:

$$w_{ij} = \begin{cases} \omega, & j = c \\ 0, & j \neq c \end{cases} \tag{5–26}$$

ω is the preset weight for that classifier, and c is the classification result of that classifier. The final decision criterion during classification is as follows:

$$R_j = \sum\nolimits_{i=1}^{n} w_{ij}, \quad j = 1, 2, \cdots, m \tag{5–27}$$

Given a recognition threshold t, if $R_j \geq t$ is satisfied, the corresponding class is output.

5.3.2.2 Combined Classifier Design Flow

Three SVM classifiers are used for classifying EEG data (after dimensionality reduction), EEG data (alpha band), and eye movement data, abbreviated as C, C_2, C_3. The design flow is as follows:

(1) C_1, C_2, C_3 learn separately in the training set and undergo parameter optimization to obtain the parameters for each classifier.

(2) Compare the recognition results of each classifier with the results in the test set and rank the reliability of each classifier based on its accuracy.

(3) Collect the results of C_1, C_2, C_3, and set the value range for each element in the matrix.

(4) Preset a rejection threshold t based on the reliability of each classifier. Determine the classification category by judging the relationship between R_j and the threshold t.

5.3.3 Classification Result Validation and Comparison

Based on the ensemble classifier workflow described earlier, dimensionality reduction and classification are performed separately on the EEG data, eye-tracking data, and alpha-band EEG data. Once classification is complete, an appropriate fusion strategy is designed to combine the results of these three modalities, and the classification performance is compared.

5.3.3.1 EEG Data Dimensionality Reduction and Classification

The factor analysis approach was applied to reduce the dimensionality of the 80-dimensional EEG signals. The resulting component matrix is provided in Appendix Table 3, and its corresponding weight coefficient matrix appears in Appendix Table 4. The scree plot of the components is displayed in Figure 5.28.

The scree plot displays the amount of information contained in each component in descending order, providing an intuitive reflection of the general situation of the components. Specific details can be viewed in Table 5.15.

It can be seen that the first five components can explain 92.269% of the variance, already containing most of the information in the entire

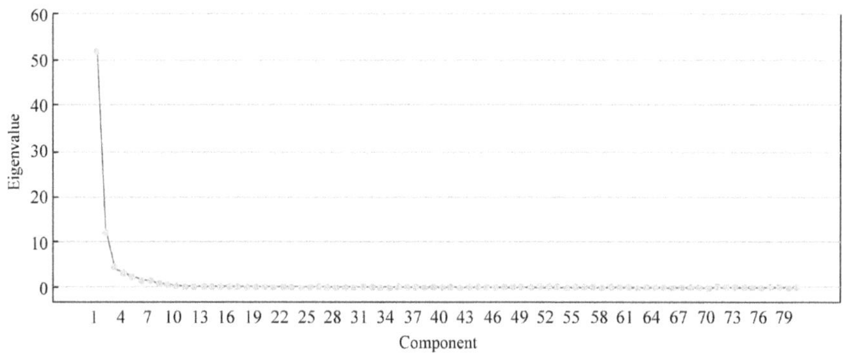

FIGURE 5.28 Scree plot for dimensionality reduction of 80-dimensional EEG data

TABLE 5.15 Information Related to Components

Component	Eigenvalue	Variance Percentage (%)	Cumulative Percentage (%)
1	51.902	64.877	64.877
2	12.025	15.032	79.909
3	4.578	5.722	85.631
4	3.018	3.773	89.404
5	2.292	2.865	92.269

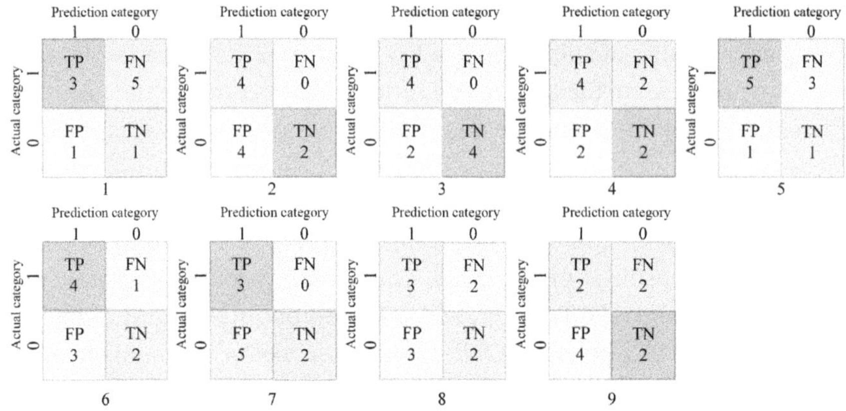

FIGURE 5.29 Confusion matrices for each subset of the 80-dimensional EEG classifier

80-dimensional EEG data. Therefore, the first five components after dimensionality reduction are taken as the main components. After dimensionality reduction is completed, an SVM classifier is used to classify the data. Considering the limited sample size (90 sets), K-fold CV ($K = 9$) is used to comprehensively evaluate the model. After the model is trained, the confusion matrices for each subset are shown in Figure 5.29, the classification performance is shown in Table 5.16, and the ROC curves are shown in Figure 5.30.

From the metrics in Table 5.16, it is suggested that the classification performance of the 80-dimensional EEG data after dimensionality reduction is not ideal. Compared to the classification performance after dimensionality reduction of the 107-dimensional EEG and eye movement data, there is still room for improvement. The potential reason is that, compared to the 107-dimensional data, the 80-dimensional EEG data lacks the eye movement data and alpha wave EEG data, which have better significance;

TABLE 5.16 Classification Performance for Each Subset of the 80-Dimensional EEG Classifier

Evaluation Matrices	Subset 1	2	3	4	5	6	7	8	9	Average
Error rate	0.6	0.4	0.2	0.4	0.4	0.4	0.5	0.5	0.6	0.44
Accuracy	0.4	0.6	0.8	0.6	0.6	0.6	0.5	0.5	0.4	0.56
Precision	0.375	1	1	0.67	0.625	0.8	1	0.6	0.5	0.73
Recall	0.75	0.5	0.67	0.67	0.83	0.57	0.375	0.5	0.33	0.58
F-score	0.5	0.67	0.8	0.67	0.714	0.67	0.55	0.55	0.4	0.61
AUC value	0.86	0.75	0.81	0.75	0.68	0.54	0.53	0.5	0.72	0.63

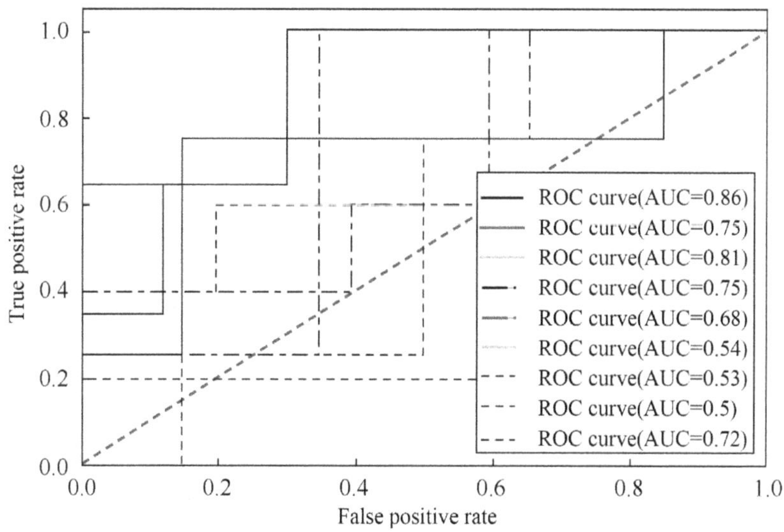

FIGURE 5.30 ROC curves for each subset of the 80-dimensional EEG classifier

hence, the suboptimal classification performance. After analyzing the model's classification performance, all sample values are used as training data to obtain the hyperplane parameters for the SVM model, specifically as, $w = [-1.681 \quad -3.181 \quad -0.032 \quad -1.475 \quad 3.276]^{\mathrm{T}}$, and $b = [1.018]$. These parameters, combined with the relationship between the decision equation $f(x) = w \cdot x + b$ and 0, can determine the category.

5.3.3.2 Eye Movement Data Classification

The SVM classifier is used to classify the eye movement data. Considering the sample size of 90 sets, *K*-fold CV ($K = 9$) is similarly used to comprehensively evaluate the model. After the model is trained, the confusion matrices for each subset are shown in Figure 5.31, the classification performance is shown in Table 5.17, and the ROC curves are shown in Figure 5.32.

Based on the evaluation matrices mentioned earlier, the classification model using eye movement data demonstrates strong classification performance. After evaluating the model's performance, all available samples were used as training data to derive the hyperplane parameters for the SVM classifier, as $w = [-3.948\ \ -0.041\ 0.355\ \ -0.461\ 1.155\ 0.998\ \ -3.521]^T$ and $b = [1.882]$.

5.3.3.3 *Alpha Band EEG Data Classification*

The 20-dimensional EEG data are classified using an SVM classifier, and *K*-fold CV (K = 9) is used to comprehensively evaluate the model. After

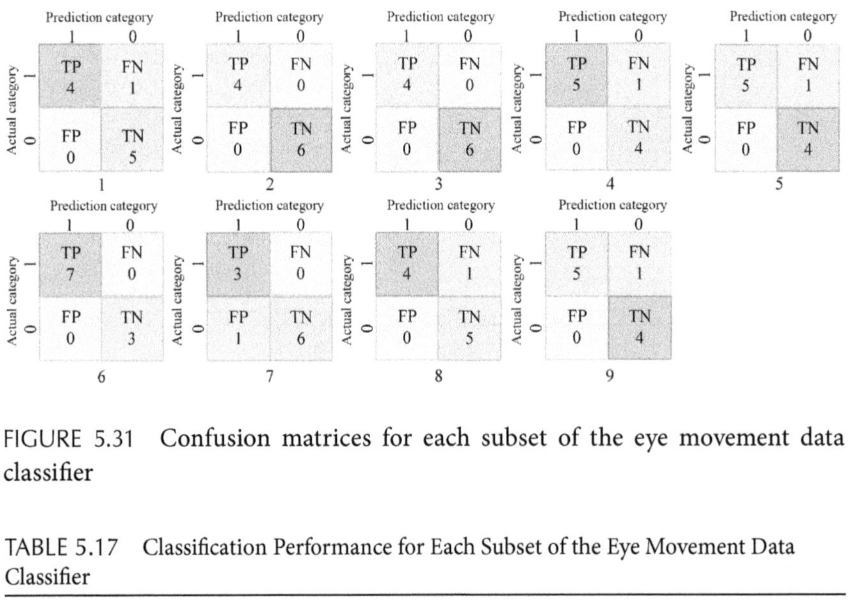

FIGURE 5.31 Confusion matrices for each subset of the eye movement data classifier

TABLE 5.17 Classification Performance for Each Subset of the Eye Movement Data Classifier

Evaluation Matrices	Subset 1	2	3	4	5	6	7	8	9	Average
Error rate	0.1	0	0	0.1	0.1	0	0.1	0.1	0.1	0.067
Accuracy	0.9	1	1	0.9	0.9	1	0.9	0.9	0.9	0.933
Precision	0.8	1	1	0.83	0.83	1	0.8	1	0.83	0.899
Recall	1	1	1	1	1	1	1	0.75	1	0.972
F-score	0.89	1	1	0.91	0.91	1	0.89	0.86	0.91	0.930
AUC value	0.88	1	1	0.9	0.88	1	0.88	0.83	0.92	0.921

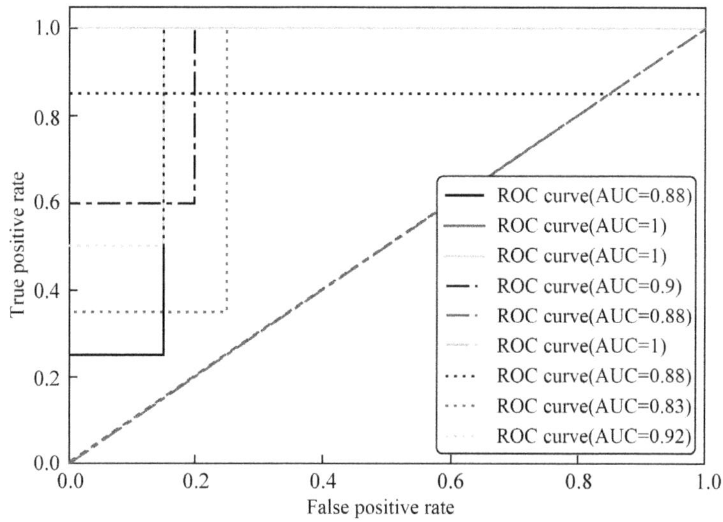

FIGURE 5.32 ROC curves for each subset of the eye movement data classifier

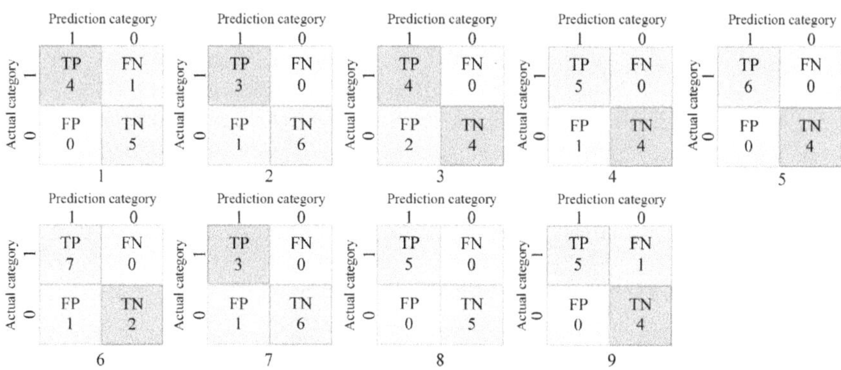

FIGURE 5.33 Confusion matrices for each subset of the alpha band EEG data classifier

the model is trained, the confusion matrices for each subset are shown in Figure 5.33, the classification performance is shown in Table 5.18, and the ROC curves are shown in Figure 5.34.

It can be seen that the 20-dimensional alpha band EEG data have good classification performance. w is a 20×1 dimensional matrix. Combined with the relationship between the decision equation $f(x) = w \cdot x + b$ and 0, the category can be determined.

TABLE 5.18 Classification Performance for Each Subset of the Alpha Band EEG Data Classifier

Evaluation Matrices	Subset 1	2	3	4	5	6	7	8	9	Average
Error rate	0.1	0.1	0.2	0.1	0	0.1	0.1	0	0.1	0.089
Accuracy	0.9	0.9	0.8	0.9	1	0.9	0.9	1	0.9	0.911
Precision	1	0.75	0.67	0.83	1	0.875	0.75	1	1	0.875
Recall	0.8	1	1	1	1	1	1	1	0.83	0.959
F-score	0.89	0.857	0.8	0.91	1	0.93	0.86	1	0.91	0.906
AUC value	1	0.95	0.96	1	1	0.94	1	1	0.95	0.978

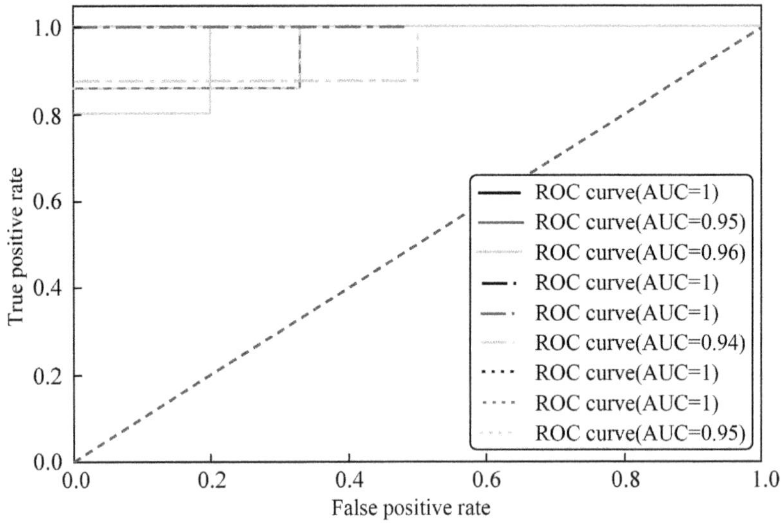

FIGURE 5.34 ROC curves for the alpha band EEG data classifier

$$w = [-3.186, -3.186,\ 1.243,\ 8.424, -3.432, -1.122, -2.386,\\ 1.913, -1.638, -9.523, 9.669, -1.049, -7.967, -6.736,\\ 1.684,\ 5.207, -9.564, -1.848,\ 1.580, -1.497]^{T} \quad (5\text{–}28)$$

$$b = [0.742884] \quad (5\text{–}29)$$

After obtaining the cognitive load SVM classification models for the three different modalities of data, appropriate combination rules are designed to combine the results of the three SVM classification models, yielding the final classification result.

5.3.3.4 Result Combination

From the classification performance evaluation metrics of the three classifiers mentioned earlier, it is suggested that the SVM classifiers based on eye movement data and EEG data (Alpha band) have good classification performance, while the SVM classifier based on EEG data (after dimensionality reduction) has poor classification performance. Therefore, when designing the combination strategy, the SVM classifiers based on eye movement and EEG data should be assigned larger weights, whereas the SVM classifier based on EEG (80-dimensional) should be assigned a smaller weight.

Assume: The SVM classifier for EEG (after dimensionality reduction), the SVM classifier for EEG (Alpha band), and the SVM for eye movement are abbreviated as C_1, C_2, and C_3, respectively.

In the combination rule, there are 3 classifiers and 2 classification outcomes. The weight matrix for classifier recognition is denoted as follows:

$$w = \begin{bmatrix} w_{11} & w_{12} \\ w_{21} & w_{22} \\ w_{31} & w_{32} \end{bmatrix} \tag{5-30}$$

Based on the classification performance of each classifier, set the value range for each element in the matrix as follows:

$$w_{1c} = \begin{cases} 1, & c = j \\ 0, & c \neq j \end{cases} \tag{5-31}$$

$$w_{2c} = \begin{cases} 1.5, & c = j \\ 0, & c \neq j \end{cases} \tag{5-32}$$

$$w_{3c} = \begin{cases} 1.5, & c = j \\ 0, & c \neq j \end{cases} \tag{5-33}$$

Here, $c = \{1, 2\}$ represents the two classification outcomes, namely high and low cognitive load, and j is the actual classification result. Calculate the value of:

$$R_j = \sum_{i=1}^{3} w_{ij}, \quad j = 1, 2 \tag{5-34}$$

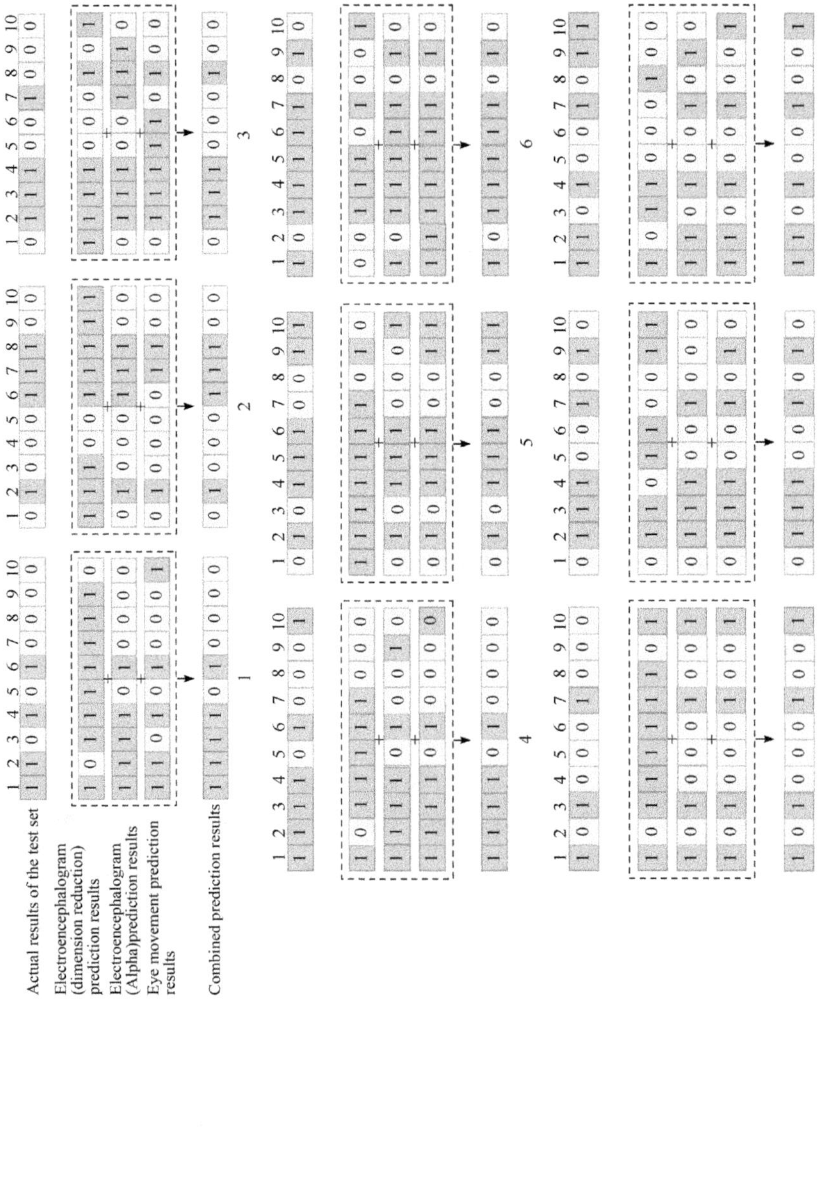

FIGURE 5.35 Combined classification results for each subset

Given the recognition threshold t, if it $R_j \geq t$ is satisfied, output the corresponding classification category. Accordingly, the combined classification results for each subset are shown in Figure 5.35, the confusion matrices for each subset are shown in Figure 5.36, the classification performance is shown in Table 5.19, and a comparative analysis of classifier performance is shown in Table 5.20.

It can be seen that the classification performance of the combined classifier has significantly improved compared to the model proposed in Chapter 3, with all positive metrics increasing by over 20%. Furthermore, the combined classifier also shows a certain degree of improvement compared to the three individual sub-classifiers.

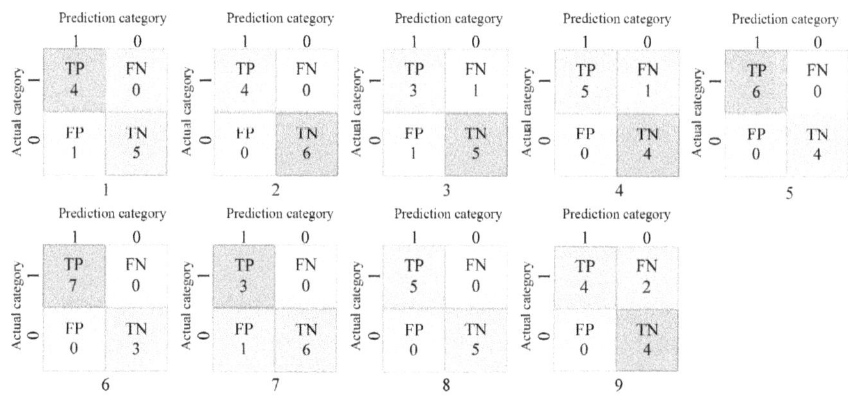

FIGURE 5.36 Confusion matrix

TABLE 5.19 Combined Classifier Classification Performance

Evaluation Matrices	Subset 1	2	3	4	5	6	7	8	9	Average
Error rate	0.1	0	0.2	0.1	0	0	0.1	0	0.2	0.078
Accuracy	0.9	1	0.8	0.9	1	1	0.9	1	0.8	0.922
Precision	0.8	1	0.75	1	1	1	0.75	1	1	0.922
Recall	1	1	0.75	0.83	1	1	1	1	0.67	0.917
F-score	0.89	1	1	0.91	1	1	0.86	1	0.8	0.940

TABLE 5.20 Comparative Analysis of Classifier Performance

Classification Data	Error Rate	Accuracy	Precision	Recall	*F*-score
EEG + eye movement (reduced dim)	0.286	0.714	0.749	0.763	0.746
EEG (reduced dim)	0.44	0.56	0.73	0.58	0.61
Eye movement data	0.067	0.933	0.899	0.972	0.9301
EEG (alpha)	0.089	0.911	0.875	0.959	0.906
Result combination	0.078	0.922	0.922	0.917	0.940

5.4 CHAPTER SUMMARY

Based on the analysis of the concepts of mental workload and cognitive workload, corresponding experimental content was designed, assessment methods were proposed, and further optimization work was conducted. The main work completed in this chapter includes the following points:

(1) Designed a simulated flight experiment capable of distinguishing between high and low mental workload levels in subjects and proposed a workload assessment method based on eye movement and physiological indicators.

(2) Designed an N-back cognitive experimental task, which is capable of distinguishing between high and low cognitive load levels in subjects, and proposed a cognitive load assessment method based on eye movement and EEG indicators.

(3) Addressed the unsatisfactory classification performance for cognitive load by employing a combined classifier approach to improve the method and optimize the model.

Appendix

APPENDIX TABLE 1 The Component Matrix of 100-Dimensional Electroencephalogram (EEG) and 7-Dimensional Eye Movement Data

	1	2	3	4	5	6	7	8	9	10	11
Delta_1	0.946	−0.139	−0.033	−0.002	−0.082	0.023	0.243	−0.061	0.076	−0.012	0.021
Delta_2	0.972	0.017	−0.080	−0.079	−0.048	0.095	0.059	−0.019	−0.020	−0.021	−0.038
Delta_3	0.787	−0.579	0.002	0.078	−0.018	0.023	−0.084	−0.020	−0.095	0.096	−0.049
Delta_4	0.835	−0.451	0.052	−0.001	−0.041	−0.051	−0.071	0.020	−0.089	0.009	−0.023
Delta_5	0.864	0.329	−0.077	−0.111	−0.051	−0.053	0.270	−0.033	0.089	0.030	0.054
Delta_6	0.671	0.632	−0.213	−0.168	0.033	0.158	−0.024	−0.007	−0.093	−0.069	−0.029
Delta_7	0.799	0.468	−0.080	−0.142	0.025	−0.103	0.100	−0.011	−0.009	0.081	0.030
Delta_8	0.968	−0.069	−0.096	0.019	−0.040	0.052	0.012	−0.050	−0.074	−0.077	−0.041
Delta_9	0.747	0.584	−0.162	−0.155	0.014	0.109	0.078	−0.024	−0.057	−0.056	0.022
Delta_10	0.272	0.215	−0.032	0.388	0.513	0.339	−0.078	−0.366	0.009	0.251	0.296
Delta_11	0.899	0.343	−0.160	−0.106	0.001	0.124	0.015	−0.023	−0.071	−0.027	−0.046
Delta_12	0.813	0.473	−0.179	−0.123	0.020	0.157	−0.085	0.004	−0.080	−0.022	−0.096
Delta_13	0.918	−0.307	0.108	0.021	0.009	−0.036	−0.139	−0.024	−0.102	0.016	0.019
Delta_14	0.712	0.644	−0.151	−0.159	−0.011	0.055	0.123	−0.012	−0.002	−0.055	0.023

(Continued)

APPENDIX TABLE 1 *(Continued)* The Component Matrix of 100-Dimensional Electroencephalogram (EEG) and 7-Dimensional Eye Movement Data

	1	2	3	4	5	6	7	8	9	10	11
Delta_15	0.967	−0.103	−0.055	−0.012	−0.067	0.016	0.188	−0.051	0.033	0.012	0.004
Delta_16	0.961	0.026	−0.001	−0.017	−0.070	−0.018	0.178	−0.070	0.070	0.003	0.066
Delta_17	0.905	−0.396	0.020	0.061	−0.046	0.064	−0.011	−0.052	−0.022	−0.033	−0.011
Delta_18	0.949	−0.018	−0.016	−0.062	−0.021	−0.059	0.159	−0.019	0.027	0.107	0.014
Delta_19	0.750	−0.015	0.029	0.316	0.469	−0.116	0.159	0.165	0.053	0.023	−0.106
Delta_20	0.903	−0.268	0.232	0.028	−0.045	−0.006	−0.062	−0.082	0.030	−0.101	0.064
Theta_1	0.942	−0.105	−0.029	0.020	−0.106	−0.059	0.220	−0.003	0.146	−0.001	0.046
Theta_2	0.547	−0.055	0.310	−0.329	−0.063	−0.023	−0.293	0.337	0.328	0.277	−0.081
Theta_3	0.755	−0.413	0.048	0.085	0.034	−0.154	−0.332	0.048	−0.157	0.182	−0.039
Theta_4	0.285	−0.071	0.514	−0.181	−0.074	−0.296	−0.255	0.417	0.249	0.206	0.104
Theta_5	0.860	0.177	0.058	−0.177	−0.088	−0.147	0.116	0.165	0.266	0.135	0.082
Theta_6	0.621	0.679	−0.078	−0.172	−0.023	0.047	−0.179	0.085	0.176	0.008	−0.037
Theta_7	0.772	0.452	0.006	−0.073	−0.017	−0.271	−0.059	0.091	0.132	0.091	0.053
Theta_8	0.749	0.093	0.134	0.049	−0.061	−0.097	−0.350	0.134	0.321	−0.029	−0.077
Theta_9	0.716	0.637	−0.142	−0.150	0.013	0.047	0.027	−0.013	−0.032	−0.036	0.003
Theta_10	0.422	0.411	−0.046	0.410	0.627	0.187	−0.077	−0.067	0.035	0.104	0.034
Theta_11	0.812	0.493	−0.151	−0.113	0.014	0.016	−0.190	0.038	−0.001	0.021	−0.054
Theta_12	0.734	0.543	−0.154	−0.108	0.028	0.074	−0.236	0.053	−0.036	0.025	−0.123
Theta_13	0.813	−0.057	0.186	0.048	0.126	−0.209	−0.389	0.016	−0.126	0.066	−0.001
Theta_14	0.635	0.718	−0.018	−0.111	−0.046	−0.105	0.017	0.080	0.085	−0.065	0.054
Theta_15	0.957	−0.010	−0.004	−0.005	−0.080	−0.146	−0.023	0.068	0.179	0.070	0.029
Theta_16	0.900	0.186	0.092	−0.033	−0.041	−0.193	0.110	0.079	0.201	0.083	0.069
Theta_17	0.934	−0.288	0.029	0.079	0.001	0.021	−0.096	−0.036	0.010	−0.040	−0.035

Theta_18	0.909	0.117	−0.004	−0.029	−0.012	−0.163	0.021	0.092	0.078	0.158	0.027
Theta_19	0.644	0.113	0.049	0.377	0.518	−0.213	0.058	0.270	0.052	0.023	−0.106
Theta_20	0.807	−0.026	0.471	0.023	0.066	−0.098	−0.149	−0.037	0.095	−0.136	0.047
Alpha_1	0.041	0.347	0.538	0.438	−0.208	0.054	−0.073	0.003	−0.015	−0.179	−0.085
Alpha_2	−0.054	0.115	0.938	0.014	−0.087	0.231	0.023	−0.029	0.047	0.103	−0.084
Alpha_3	0.125	0.060	0.732	0.015	0.015	−0.115	−0.283	−0.057	0.096	−0.133	0.354
Alpha_4	0.258	−0.311	−0.073	−0.354	0.192	−0.380	−0.191	−0.013	0.044	0.025	0.388
Alpha_5	0.078	0.435	0.488	0.510	−0.289	−0.021	0.180	0.218	−0.229	−0.034	0.026
Alpha_6	0.016	0.059	0.849	−0.389	0.199	−0.055	0.144	−0.112	−0.004	−0.034	−0.092
Alpha_7	0.186	0.506	0.463	0.551	−0.223	−0.140	0.050	0.053	−0.093	−0.228	0.072
Alpha_8	0.047	0.243	0.566	0.310	−0.186	0.211	−0.130	−0.293	0.326	0.033	−0.249
Alpha_9	0.004	0.026	0.833	−0.416	0.205	−0.038	0.185	−0.084	−0.088	−0.024	−0.104
Alpha_10	−0.007	−0.051	0.684	−0.506	0.320	−0.025	0.148	−0.184	0.018	−0.054	−0.185
Alpha_11	−0.027	0.056	0.901	−0.103	0.066	0.158	−0.027	−0.142	0.105	−0.038	−0.152
Alpha_12	0.001	−0.039	0.789	−0.476	0.255	−0.002	0.083	−0.140	−0.021	0.009	−0.110
Alpha_13	0.097	0.060	0.734	−0.373	0.181	−0.207	−0.020	−0.160	0.017	−0.107	0.097
Alpha_14	0.028	0.149	0.961	−0.125	0.015	−0.007	0.029	−0.068	0.038	−0.044	0.054
Alpha_15	−0.054	0.304	0.587	0.538	−0.354	0.259	−0.064	0.060	0.113	−0.035	−0.009
Alpha_16	−0.150	0.176	0.659	0.363	−0.309	0.401	0.074	0.192	−0.057	0.199	−0.022
Alpha_17	−0.022	−0.112	0.607	−0.520	0.256	0.032	0.202	−0.086	−0.125	0.105	−0.027
Alpha_18	0.092	0.419	0.577	0.514	−0.244	0.014	−0.045	0.084	−0.039	−0.127	0.198
Alpha_19	−0.174	0.147	0.539	0.402	−0.337	0.479	−0.001	0.203	0.008	0.248	−0.011
Alpha_20	−0.037	0.043	0.817	−0.295	0.067	0.055	0.173	0.014	−0.167	0.162	−0.029
Beta_1	0.870	−0.420	−0.022	0.070	−0.099	0.074	0.169	−0.057	0.047	−0.041	0.000

(Continued)

APPENDIX TABLE 1 *(Continued)* The Component Matrix of 100-Dimensional Electroencephalogram (EEG) and 7-Dimensional Eye Movement Data

	1	**2**	**3**	**4**	**5**	**6**	**7**	**8**	**9**	**10**	**11**
Beta_2	0.941	−0.250	−0.057	0.037	−0.089	0.074	0.153	−0.035	0.028	−0.010	−0.002
Beta_3	0.782	−0.600	0.011	0.099	−0.065	0.060	0.041	−0.044	−0.022	0.018	−0.031
Beta_4	0.815	−0.545	−0.004	0.087	−0.064	0.093	0.058	−0.049	−0.021	−0.020	−0.033
Beta_5	0.933	0.082	−0.110	−0.013	−0.099	−0.011	0.257	−0.016	0.084	−0.001	0.058
Beta_6	0.765	0.555	−0.207	−0.132	0.006	0.097	0.073	0.013	−0.073	−0.012	0.014
Beta_7	0.928	0.214	−0.117	−0.053	−0.038	−0.048	0.112	0.013	0.006	0.052	0.039
Beta_8	0.892	−0.417	0.009	0.063	−0.066	0.067	0.068	−0.056	−0.008	−0.024	−0.029
Beta_9	0.765	0.533	−0.142	−0.111	−0.031	0.021	0.201	0.001	0.017	0.006	0.110
Beta_10	0.094	0.078	0.032	0.590	0.767	0.068	0.114	−0.001	0.076	0.037	0.029
Beta_11	0.974	0.038	−0.119	−0.022	−0.062	0.088	0.129	−0.022	−0.008	−0.010	−0.004
Beta_12	0.879	0.379	−0.166	−0.090	−0.024	0.138	0.038	0.009	−0.037	−0.034	−0.040
Beta_13	0.893	−0.391	0.148	0.065	−0.056	−0.027	0.051	−0.028	−0.030	−0.061	0.058
Beta_14	0.794	0.492	−0.155	−0.107	−0.055	0.013	0.230	0.015	0.047	−0.008	0.074
Beta_15	0.905	−0.350	−0.043	0.056	−0.092	0.066	0.165	−0.050	0.028	−0.026	−0.004
Beta_16	0.936	−0.266	0.061	0.047	−0.089	0.011	0.138	−0.071	0.048	−0.061	0.048
Beta_17	0.879	−0.430	0.074	0.089	−0.075	0.018	0.071	−0.049	0.008	−0.072	0.035
Beta_18	0.931	−0.301	−0.041	0.036	−0.074	0.030	0.135	−0.021	0.029	0.013	0.017
Beta_19	0.466	−0.101	0.041	0.489	0.612	−0.090	0.194	0.230	0.087	−0.093	−0.149
Beta_20	0.870	−0.359	0.216	0.091	−0.078	−0.050	0.030	−0.064	0.035	−0.132	0.091
Gamma_1	0.781	−0.568	0.003	0.098	−0.092	0.096	0.125	−0.063	0.026	−0.042	−0.019
Gamma_2	0.888	−0.226	0.008	0.033	−0.060	0.162	−0.109	−0.031	0.004	0.005	−0.144
Gamma_3	0.816	−0.382	0.014	0.040	0.048	−0.026	−0.272	0.017	−0.197	0.171	−0.081
Gamma_4	0.831	−0.463	0.009	0.025	0.020	0.097	−0.088	−0.010	−0.138	0.053	−0.114

Gamma_5	0.929	−0.259	−0.040	0.031	−0.080	0.089	0.135	−0.053	0.010	−0.047	−0.003
Gamma_6	0.712	0.463	−0.145	−0.133	0.101	0.165	−0.334	0.034	−0.232	0.038	−0.134
Gamma_7	0.937	−0.142	−0.066	−0.003	0.011	0.011	−0.116	−0.007	−0.122	0.088	−0.049
Gamma_8	0.794	−0.470	0.055	0.080	−0.052	0.115	−0.124	−0.012	−0.019	0.067	−0.152
Gamma_9	0.570	0.610	0.068	−0.093	0.125	0.181	−0.110	−0.188	−0.030	0.030	0.252
Gamma_10	−0.045	0.041	0.020	0.269	0.245	0.398	−0.098	−0.553	0.064	0.340	0.430
Gamma_11	0.822	0.372	−0.080	−0.110	0.061	0.148	−0.274	0.005	−0.174	0.028	−0.119
Gamma_12	0.554	0.631	−0.148	−0.131	0.031	0.269	−0.205	0.036	−0.080	−0.084	−0.154
Gamma_13	0.807	−0.152	0.347	0.048	0.076	−0.163	−0.201	0.078	−0.277	−0.010	0.075
Gamma_14	0.678	0.634	−0.120	−0.180	0.031	0.160	0.014	−0.011	−0.085	−0.120	0.010
Gamma_15	0.906	−0.350	−0.043	0.051	−0.070	0.116	0.100	−0.050	−0.021	−0.049	−0.024
Gamma_16	0.928	−0.324	0.072	0.060	−0.036	0.036	0.004	−0.049	−0.035	−0.052	0.000
Gamma_17	0.884	−0.153	0.243	0.089	0.049	−0.127	−0.199	0.008	−0.165	−0.096	0.056
Gamma_18	0.856	−0.118	−0.089	−0.054	0.055	0.047	−0.123	0.042	−0.107	0.101	−0.012
Gamma_19	0.085	0.029	0.047	0.534	0.719	−0.108	0.152	0.267	0.091	−0.110	−0.179
Gamma_20	0.809	−0.092	0.376	0.109	0.041	−0.181	−0.201	−0.013	−0.140	−0.167	0.106
Perclos	−0.129	−0.022	0.332	−0.267	0.143	0.220	0.312	0.474	−0.382	0.004	0.111
Fixation_Duration	0.076	0.350	−0.131	0.287	−0.205	−0.568	0.083	−0.423	0.020	0.216	−0.223
Number_Of_Fixations	0.002	−0.409	−0.009	−0.317	0.147	0.466	0.001	0.466	0.052	−0.184	0.299
Saccade_Duration	0.025	0.068	0.055	0.013	−0.120	−0.068	0.339	0.246	−0.086	0.575	0.062
Saccade_Angle	0.001	−0.121	−0.193	−0.261	0.136	0.392	−0.080	−0.124	0.357	−0.165	−0.263
Number_Of_Saccades	0.078	−0.484	−0.148	−0.226	0.125	0.493	−0.121	0.232	0.228	−0.291	0.210
Pupil_Area	0.264	0.479	0.185	0.299	−0.097	−0.476	0.050	−0.112	−0.106	−0.305	0.089

APPENDIX TABLE 2 The Components Weight Coefficient Matrix of 00-Dimensional Electroencephalogram (EEG) and 7-Dimensional Eye Movement Data

	1	2	3	4	5	6	7	8	9	10	11
Delta_1	0.018	−0.011	−0.003	0.000	−0.020	0.007	0.091	−0.027	0.047	−0.008	0.014
Delta_2	0.019	0.001	−0.007	−0.014	−0.012	0.028	0.022	−0.008	−0.012	−0.013	−0.026
Delta_3	0.015	−0.044	0.000	0.013	−0.004	0.007	−0.032	−0.009	−0.059	0.060	−0.033
Delta_4	0.016	−0.035	0.005	0.000	−0.010	−0.015	−0.027	0.009	−0.055	0.006	−0.016
Delta_5	0.017	0.025	−0.007	−0.019	−0.012	−0.016	0.102	−0.014	0.055	0.019	0.036
Delta_6	0.013	0.048	−0.018	−0.029	0.008	0.047	−0.009	−0.003	−0.058	−0.044	−0.020
Delta_7	0.015	0.036	−0.007	−0.024	0.006	−0.031	0.038	−0.005	−0.005	0.051	0.020
Delta_8	0.019	−0.005	−0.008	0.003	−0.010	0.016	0.005	−0.022	−0.046	−0.048	−0.027
Delta_9	0.014	0.045	−0.014	−0.027	0.003	0.033	0.030	−0.010	−0.036	−0.035	0.015
Delta_10	0.005	0.016	−0.003	0.067	0.125	0.102	−0.029	−0.160	0.005	0.158	0.200
Delta_11	0.017	0.026	−0.014	−0.018	0.000	0.037	0.005	−0.010	−0.044	−0.017	−0.031
Delta_12	0.016	0.036	−0.015	−0.021	0.005	0.047	−0.032	0.002	−0.050	−0.014	−0.065
Delta_13	0.018	−0.024	0.009	0.004	0.002	−0.011	−0.052	−0.011	−0.064	0.010	0.013
Delta_14	0.014	0.049	−0.013	−0.027	−0.003	0.016	0.046	−0.005	−0.001	−0.034	0.015
Delta_15	0.019	−0.008	−0.005	−0.002	−0.016	0.005	0.071	−0.022	0.020	0.007	0.003
Delta_16	0.018	0.002	0.000	−0.003	−0.017	−0.005	0.067	−0.030	0.044	0.002	0.045
Delta_17	0.017	−0.030	0.002	0.010	−0.011	0.019	−0.004	−0.023	−0.014	−0.021	−0.007
Delta_18	0.018	−0.001	−0.001	−0.011	−0.005	−0.018	0.060	−0.008	0.017	0.068	0.009
Delta_19	0.014	−0.001	0.003	0.054	0.115	−0.035	0.060	0.072	0.033	0.014	−0.072
Delta_20	0.017	−0.021	0.020	0.005	−0.011	−0.002	−0.023	−0.036	0.019	−0.063	0.043
Theta_1	0.018	−0.008	−0.003	0.003	−0.026	−0.018	0.083	−0.001	0.091	−0.001	0.031
Theta_2	0.010	−0.004	0.027	−0.057	−0.015	−0.007	−0.110	0.147	0.204	0.174	−0.055
Theta_3	0.014	−0.032	0.004	0.015	0.008	−0.046	−0.125	0.021	−0.098	0.114	−0.026

Theta_4	0.005	−0.005	0.044	−0.031	−0.018	−0.089	−0.096	0.182	0.155	0.130	0.070
Theta_5	0.016	0.014	0.005	−0.030	−0.022	−0.044	0.044	0.072	0.166	0.085	0.055
Theta_6	0.012	0.052	−0.007	−0.030	−0.006	0.014	−0.067	0.037	0.110	0.005	−0.025
Theta_7	0.015	0.035	0.000	−0.013	−0.004	−0.081	−0.022	0.040	0.082	0.057	0.036
Theta_8	0.014	0.007	0.012	0.008	−0.015	−0.029	−0.132	0.059	0.200	−0.018	−0.052
Theta_9	0.014	0.049	−0.012	−0.026	0.003	0.014	0.010	−0.006	−0.020	−0.023	0.002
Theta_10	0.008	0.031	−0.004	0.071	0.153	0.056	−0.029	−0.029	0.022	0.065	0.023
Theta_11	0.016	0.038	−0.013	−0.019	0.004	0.005	−0.072	0.016	0.000	0.013	−0.037
Theta_12	0.014	0.042	−0.013	−0.019	0.007	0.022	−0.089	0.023	−0.022	0.016	−0.083
Theta_13	0.016	−0.004	0.016	0.008	0.031	−0.063	−0.147	0.007	−0.078	0.041	0.000
Theta_14	0.012	0.055	−0.002	−0.019	−0.011	−0.032	0.006	0.035	0.053	−0.041	0.037
Theta_15	0.018	−0.001	0.000	−0.001	−0.020	−0.044	−0.009	0.030	0.111	0.044	0.020
Theta_16	0.017	0.014	0.008	−0.006	−0.010	−0.058	0.041	0.035	0.125	0.052	0.046
Theta_17	0.018	−0.022	0.002	0.014	0.000	0.006	−0.036	−0.016	0.006	−0.025	−0.024
Theta_18	0.017	0.009	0.000	−0.005	−0.003	−0.049	0.008	0.040	0.048	0.099	0.018
Theta_19	0.012	0.009	0.004	0.065	0.127	−0.064	0.022	0.118	0.032	0.014	−0.072
Theta_20	0.015	−0.002	0.041	0.004	0.016	−0.030	−0.056	−0.016	0.059	−0.085	0.032
Alpha_1	0.001	0.027	0.046	0.075	−0.051	0.016	−0.028	0.001	−0.009	−0.112	−0.058
Alpha_2	−0.001	0.009	0.081	0.002	−0.021	0.069	0.009	−0.012	0.029	0.065	−0.057
Alpha_3	0.002	0.005	0.063	0.003	0.004	−0.035	−0.107	−0.025	0.060	−0.084	0.239
Alpha_4	0.005	−0.024	−0.006	−0.061	0.047	−0.114	−0.072	−0.005	0.027	0.016	0.263
Alpha_5	0.002	0.033	0.042	0.088	−0.071	−0.006	0.068	0.095	−0.142	−0.021	0.018
Alpha_6	0.000	0.005	0.073	−0.067	0.049	−0.016	0.054	−0.049	−0.002	−0.022	−0.062
Alpha_7	0.004	0.039	0.040	0.095	−0.055	−0.042	0.019	0.023	−0.058	−0.143	0.049

(Continued)

APPENDIX TABLE 2 *(Continued)* The Components Weight Coefficient Matrix of 00-Dimensional Electroencephalogram (EEG) and 7-Dimensional Eye Movement Data

	1	**2**	**3**	**4**	**5**	**6**	**7**	**8**	**9**	**10**	**11**
Alpha_8	0.001	0.019	0.049	0.053	−0.045	0.063	−0.049	−0.128	0.202	0.021	−0.168
Alpha_9	0.000	0.002	0.072	−0.072	0.050	−0.011	0.070	−0.036	−0.054	−0.015	−0.070
Alpha_10	0.000	−0.004	0.059	−0.087	0.078	−0.007	0.056	−0.080	0.011	−0.034	−0.125
Alpha_11	−0.001	0.004	0.078	−0.018	0.016	0.048	−0.010	−0.062	0.065	−0.024	−0.103
Alpha_12	0.000	−0.003	0.068	−0.082	0.062	0.000	0.031	−0.061	−0.013	0.006	−0.074
Alpha_13	0.002	0.005	0.063	−0.064	0.044	−0.062	−0.007	−0.070	0.011	−0.067	0.066
Alpha_14	0.001	0.011	0.083	−0.022	0.004	−0.002	0.011	−0.030	0.024	−0.028	0.036
Alpha_15	−0.001	0.023	0.051	0.093	−0.087	0.078	−0.024	0.026	0.070	−0.022	−0.006
Alpha_16	−0.003	0.013	0.057	0.062	−0.075	0.120	0.028	0.084	−0.035	0.125	−0.015
Alpha_17	0.000	−0.009	0.052	−0.089	0.062	0.010	0.076	−0.037	−0.078	0.066	−0.019
Alpha_18	0.002	0.032	0.050	0.088	−0.060	0.004	−0.017	0.037	−0.024	−0.080	0.134
Alpha_19	−0.003	0.011	0.047	0.069	−0.082	0.144	0.000	0.089	0.005	0.156	−0.007
Alpha_20	−0.001	0.003	0.071	−0.051	0.016	0.016	0.065	0.006	−0.104	0.102	−0.019
Beta_1	0.017	−0.032	−0.002	0.012	−0.024	0.022	0.064	−0.025	0.029	−0.026	0.000
Beta_2	0.018	−0.019	−0.005	0.006	−0.022	0.022	0.058	−0.015	0.018	−0.006	−0.001
Beta_3	0.015	−0.046	0.001	0.017	−0.016	0.018	0.016	−0.019	−0.014	0.012	−0.021
Beta_4	0.016	−0.042	0.000	0.015	−0.016	0.028	0.022	−0.021	−0.013	−0.013	−0.022
Beta_5	0.018	0.006	−0.010	−0.002	−0.024	−0.003	0.097	−0.007	0.052	0.000	0.039
Beta_6	0.015	0.042	−0.018	−0.023	0.001	0.029	0.028	0.006	−0.045	−0.007	0.009
Beta_7	0.018	0.016	−0.010	−0.009	−0.009	−0.014	0.042	0.006	0.003	0.033	0.027
Beta_8	0.017	−0.032	0.001	0.011	−0.016	0.020	0.026	−0.024	−0.005	−0.015	−0.020
Beta_9	0.015	0.041	−0.012	−0.019	−0.008	0.006	0.076	0.001	0.010	0.003	0.074

Beta_10	0.002	0.006	0.003	0.102	0.187	0.020	0.043	0.000	0.047	0.023	0.020
Beta_11	0.019	0.003	−0.010	−0.004	−0.015	0.026	0.048	−0.010	−0.005	−0.006	−0.002
Beta_12	0.017	0.029	−0.014	−0.015	−0.006	0.041	0.014	0.004	−0.023	−0.022	−0.027
Beta_13	0.017	−0.030	0.013	0.011	−0.014	−0.008	0.019	−0.012	−0.019	−0.038	0.039
Beta_14	0.015	0.038	−0.013	−0.018	−0.013	0.004	0.087	0.006	0.029	−0.005	0.050
Beta_15	0.017	−0.027	−0.004	0.010	−0.023	0.020	0.062	−0.022	0.017	−0.016	−0.003
Beta_16	0.018	−0.020	0.005	0.008	−0.022	0.003	0.052	−0.031	0.030	−0.039	0.032
Beta_17	0.017	−0.033	0.006	0.015	−0.018	0.006	0.027	−0.021	0.005	−0.045	0.023
Beta_18	0.018	−0.023	−0.004	0.006	−0.018	0.009	0.051	−0.009	0.018	0.008	0.012
Beta_19	0.009	−0.008	0.004	0.084	0.150	−0.027	0.073	0.100	0.054	−0.058	−0.101
Beta_20	0.017	−0.027	0.019	0.016	−0.019	−0.015	0.011	−0.028	0.022	−0.083	0.062
Gamma_1	0.015	−0.043	0.000	0.017	−0.023	0.029	0.047	−0.028	0.016	−0.026	−0.013
Gamma_2	0.017	−0.017	0.001	0.006	−0.015	0.049	−0.041	−0.014	0.003	0.003	−0.098
Gamma_3	0.016	−0.029	0.001	0.007	0.012	−0.008	−0.102	0.008	−0.123	0.107	−0.055
Gamma_4	0.016	−0.035	0.001	0.004	0.005	0.029	−0.033	−0.004	−0.086	0.033	−0.077
Gamma_5	0.018	−0.020	−0.003	0.005	−0.020	0.027	0.051	−0.023	0.006	−0.030	−0.002
Gamma_6	0.014	0.035	−0.012	−0.023	0.025	0.050	−0.126	0.015	−0.144	0.024	−0.090
Gamma_7	0.018	−0.011	−0.006	0.000	0.003	0.003	−0.044	−0.003	−0.076	0.055	−0.033
Gamma_8	0.015	−0.036	0.005	0.014	−0.013	0.034	−0.047	−0.005	−0.012	0.042	−0.103
Gamma_9	0.011	0.047	0.006	−0.016	0.031	0.054	−0.042	−0.082	−0.018	0.019	0.170
Gamma_10	−0.001	0.003	0.002	0.046	0.060	0.119	−0.037	−0.241	0.040	0.214	0.291
Gamma_11	0.016	0.029	−0.007	−0.019	0.015	0.044	−0.103	0.002	−0.108	0.017	−0.080
Gamma_12	0.011	0.048	−0.013	−0.023	0.007	0.081	−0.077	0.016	−0.049	−0.053	−0.104
Gamma_13	0.015	−0.012	0.030	0.008	0.019	−0.049	−0.076	0.034	−0.172	−0.007	0.051

(*Continued*)

APPENDIX TABLE 2 *(Continued)* The Components Weight Coefficient Matrix of 00-Dimensional Electroencephalogram (EEG) and 7-Dimensional Eye Movement Data

	1	2	3	4	5	6	7	8	9	10	11
Gamma_14	0.013	0.049	−0.010	−0.031	0.007	0.048	0.005	−0.005	−0.053	−0.076	0.007
Gamma_15	0.017	−0.027	−0.004	0.009	−0.017	0.035	0.038	−0.022	−0.013	−0.031	−0.016
Gamma_16	0.018	−0.025	0.006	0.010	−0.009	0.011	0.002	−0.021	−0.022	−0.033	0.000
Gamma_17	0.017	−0.012	0.021	0.015	0.012	−0.038	−0.075	0.004	−0.103	−0.061	0.038
Gamma_18	0.016	−0.009	−0.008	−0.009	0.013	0.014	−0.046	0.018	−0.066	0.064	−0.008
Gamma_19	0.002	0.002	0.004	0.092	0.176	−0.032	0.057	0.117	0.056	−0.069	−0.121
Gamma_20	0.015	−0.007	0.032	0.019	0.010	−0.054	−0.076	−0.006	−0.087	−0.105	0.072
Perclos	−0.002	−0.002	0.029	−0.046	0.035	0.066	0.118	0.207	−0.238	0.002	0.075
Fixation_Duration	0.001	0.027	−0.011	0.049	−0.050	−0.170	0.031	−0.185	0.012	0.136	−0.151
Number_Of_Fixations	0.000	−0.031	−0.001	−0.054	0.036	0.140	0.000	0.203	0.033	−0.116	0.202
Saccade_Duration	0.000	0.005	0.005	0.002	−0.029	−0.020	0.128	0.107	−0.054	0.362	0.042
Saccade_Angle	0.000	−0.009	−0.017	−0.045	0.033	0.117	−0.030	−0.054	0.222	−0.104	−0.178
Number_Of_Saccades	0.001	−0.037	−0.013	−0.039	0.031	0.148	−0.046	0.101	0.142	−0.183	0.142
Pupil_Area	0.005	0.037	0.016	0.051	−0.024	−0.143	0.019	−0.049	−0.066	−0.192	0.060

APPENDIX TABLE 3 The Data Component Matrix of 80-Dimensional Electroence-Phalogram

	1	2	3	4	5
Delta_1	0.947	−0.131	−0.069	−0.195	−0.165
Delta_2	0.973	0.049	−0.084	−0.121	0.024
Delta_3	0.790	−0.569	−0.002	−0.030	0.121
Delta_4	0.836	−0.449	−0.061	0.058	0.075
Delta_5	0.863	0.346	−0.084	−0.127	−0.258
Delta_6	0.670	0.689	−0.044	−0.111	0.110
Delta_7	0.797	0.490	−0.041	0.013	−0.149
Delta_8	0.969	−0.053	−0.029	−0.121	0.069
Delta_9	0.746	0.628	−0.050	−0.127	0.004
Delta_10	0.273	0.196	0.670	−0.242	0.387
Delta_11	0.899	0.387	−0.047	−0.124	0.074
Delta_12	0.812	0.521	−0.036	−0.068	0.158
Delta_13	0.919	−0.317	0.002	0.105	0.121
Delta_14	0.710	0.679	−0.068	−0.107	−0.078
Delta_15	0.968	−0.089	−0.063	−0.165	−0.119
Delta_16	0.961	0.025	−0.057	−0.107	−0.143
Delta_17	0.907	−0.393	−0.021	−0.080	0.072
Delta_18	0.949	−0.004	−0.051	−0.058	−0.159
Delta_19	0.750	−0.044	0.564	0.056	−0.265
Delta_20	0.903	−0.307	−0.024	0.092	0.055
Theta_1	0.942	−0.107	−0.074	−0.112	−0.236
Theta_2	0.548	−0.043	−0.230	0.556	0.018
Theta_3	0.756	−0.423	0.046	0.261	0.197
Theta_4	0.283	−0.144	−0.150	0.783	−0.118
Theta_5	0.860	0.187	−0.157	0.170	−0.303
Theta_6	0.619	0.705	−0.082	0.171	0.053
Theta_7	0.769	0.437	−0.036	0.275	−0.199
Theta_8	0.747	0.053	−0.013	0.412	0.065
Theta_9	0.714	0.671	−0.047	−0.048	−0.005
Theta_10	0.421	0.386	0.776	−0.044	0.174
Theta_11	0.810	0.529	−0.037	0.093	0.115
Theta_12	0.733	0.579	−0.021	0.098	0.190
Theta_13	0.811	−0.095	0.120	0.411	0.190
Theta_14	0.631	0.707	−0.063	0.144	−0.129
Theta_15	0.956	−0.015	−0.068	0.126	−0.143
Theta_16	0.898	0.159	−0.035	0.162	−0.268
Theta_17	0.934	−0.293	0.030	0.017	0.093
Theta_18	0.908	0.116	−0.020	0.142	−0.175
Theta_19	0.642	0.063	0.642	0.222	−0.273
Theta_20	0.804	−0.113	0.082	0.364	0.027

(Continued)

APPENDIX TABLE 3 *(Continued)* The Data Component Matrix of 80-Dimensional Electroencephalogram

	1	2	3	4	5
Beta_1	0.872	−0.415	−0.056	−0.211	−0.067
Beta_2	0.943	−0.238	−0.059	−0.191	−0.061
Beta_3	0.785	−0.598	−0.024	−0.129	0.040
Beta_4	0.818	−0.538	−0.027	−0.159	0.047
Beta_5	0.933	0.097	−0.080	−0.190	−0.226
Beta_6	0.764	0.603	−0.046	−0.129	−0.005
Beta_7	0.927	0.236	−0.053	−0.068	−0.137
Beta_8	0.894	−0.415	−0.035	−0.135	0.022
Beta_9	0.764	0.561	−0.059	−0.132	−0.164
Beta_10	0.093	0.023	0.978	−0.071	−0.057
Beta_11	0.975	0.064	−0.059	−0.180	−0.040
Beta_12	0.879	0.419	−0.053	−0.130	0.045
Beta_13	0.893	−0.419	−0.021	0.000	−0.022
Beta_14	0.793	0.521	−0.079	−0.151	−0.204
Beta_15	0.907	−0.341	−0.056	−0.205	−0.067
Beta_16	0.936	−0.280	−0.051	−0.111	−0.082
Beta_17	0.880	−0.447	−0.026	−0.084	−0.018
Beta_18	0.933	−0.289	−0.053	−0.150	−0.083
Beta_19	0.466	−0.150	0.777	0.033	−0.300
Beta_20	0.869	−0.408	−0.020	0.039	−0.036
Gamma_1	0.784	−0.565	−0.042	−0.204	−0.007
Gamma_2	0.890	−0.220	−0.035	−0.050	0.193
Gamma_3	0.818	−0.372	0.035	0.143	0.242
Gamma_4	0.834	−0.445	0.003	−0.038	0.175
Gamma_5	0.931	−0.248	−0.057	−0.191	−0.031
Gamma_6	0.712	0.514	0.018	0.094	0.375
Gamma_7	0.939	−0.120	−0.007	0.010	0.121
Gamma_8	0.797	−0.474	−0.015	−0.008	0.189
Gamma_9	0.568	0.611	0.093	0.018	0.238
Gamma_10	−0.044	0.024	0.382	−0.315	0.478
Gamma_11	0.822	0.406	−0.001	0.082	0.318
Gamma_12	0.553	0.674	−0.022	−0.016	0.299
Gamma_13	0.805	−0.221	0.086	0.366	0.128
Gamma_14	0.677	0.675	−0.048	−0.086	0.082
Gamma_15	0.908	−0.336	−0.043	−0.195	0.024
Gamma_16	0.929	−0.336	−0.007	−0.042	0.052
Gamma_17	0.882	−0.210	0.083	0.264	0.123
Gamma_18	0.858	−0.075	−0.003	0.016	0.121
Gamma_19	0.083	−0.028	0.894	0.094	−0.294
Gamma_20	0.805	−0.183	0.096	0.349	0.091

APPENDIX TABLE 4 The Component Weight Coefficient Matrix of 80-Dimensional Electroencephalogram (EEG) Data

	1	2	3	4	5
Delta_1	0.018	−0.011	−0.015	−0.065	−0.072
Delta_2	0.019	0.004	−0.018	−0.040	0.010
Delta_3	0.015	−0.047	0.000	−0.010	0.053
Delta_4	0.016	−0.037	−0.013	0.019	0.033
Delta_5	0.017	0.029	−0.018	−0.042	−0.112
Delta_6	0.013	0.057	−0.010	−0.037	0.048
Delta_7	0.015	0.041	−0.009	0.004	−0.065
Delta_8	0.019	−0.004	−0.006	−0.040	0.030
Delta_9	0.014	0.052	−0.011	−0.042	0.002
Delta_10	0.005	0.016	0.146	−0.080	0.169
Delta_11	0.017	0.032	−0.010	−0.041	0.032
Delta_12	0.016	0.043	−0.008	−0.023	0.069
Delta_13	0.018	−0.026	0.001	0.035	0.053
Delta_14	0.014	0.056	−0.015	−0.035	−0.034
Delta_15	0.019	−0.007	−0.014	−0.055	−0.052
Delta_16	0.019	0.002	−0.012	−0.035	−0.062
Delta_17	0.017	−0.033	−0.005	−0.027	0.031
Delta_18	0.018	0.000	−0.011	−0.019	−0.069
Delta_19	0.014	−0.004	0.123	0.019	−0.115
Delta_20	0.017	−0.025	−0.005	0.030	0.024
Theta_1	0.018	−0.009	−0.016	−0.037	−0.103
Theta_2	0.011	−0.004	−0.050	0.184	0.008
Theta_3	0.015	−0.035	0.010	0.086	0.086
Theta_4	0.005	−0.012	−0.033	0.259	−0.052
Theta_5	0.017	0.016	−0.034	0.056	−0.132
Theta_6	0.012	0.059	−0.018	0.057	0.023
Theta_7	0.015	0.036	−0.008	0.091	−0.087
Theta_8	0.014	0.004	−0.003	0.137	0.028
Theta_9	0.014	0.056	−0.010	−0.016	−0.002
Theta_10	0.008	0.032	0.169	−0.015	0.076
Theta_11	0.016	0.044	−0.008	0.031	0.050
Theta_12	0.014	0.048	−0.005	0.033	0.083
Theta_13	0.016	−0.008	0.026	0.136	0.083
Theta_14	0.012	0.059	−0.014	0.048	−0.056
Theta_15	0.018	−0.001	−0.015	0.042	−0.063
Theta_16	0.017	0.013	−0.008	0.054	−0.117
Theta_17	0.018	−0.024	0.007	0.006	0.041
Theta_18	0.017	0.010	−0.004	0.047	−0.076
Theta_19	0.012	0.005	0.140	0.074	−0.119

(*Continued*)

APPENDIX TABLE 4 *(Continued)* The Component Weight Coefficient Matrix of 80-Dimensional Electroencephalogram (EEG) Data

	1	2	3	4	5
Theta_20	0.015	−0.009	0.018	0.121	0.012
Beta_1	0.017	−0.034	−0.012	−0.070	−0.029
Beta_2	0.018	−0.020	−0.013	−0.063	−0.027
Beta_3	0.015	−0.050	−0.005	−0.043	0.018
Beta_4	0.016	−0.045	−0.006	−0.053	0.021
Beta_5	0.018	0.008	−0.017	−0.063	−0.099
Beta_6	0.015	0.050	−0.010	−0.043	−0.002
Beta_7	0.018	0.020	−0.012	−0.022	−0.060
Beta_8	0.017	−0.035	−0.008	−0.045	0.010
Beta_9	0.015	0.047	−0.013	−0.044	−0.072
Beta_10	0.002	0.002	0.214	−0.024	−0.025
Beta_11	0.019	0.005	−0.013	−0.060	−0.018
Beta_12	0.017	0.035	−0.012	−0.043	0.020
Beta_13	0.017	−0.035	−0.005	0.000	−0.010
Beta_14	0.015	0.043	−0.017	−0.050	−0.089
Beta_15	0.017	−0.028	−0.012	−0.068	−0.029
Beta_16	0.018	−0.023	−0.011	−0.037	−0.036
Beta_17	0.017	−0.037	−0.006	−0.028	−0.008
Beta_18	0.018	−0.024	−0.011	−0.050	−0.036
Beta_19	0.009	−0.013	0.170	0.011	−0.131
Beta_20	0.017	−0.034	−0.004	0.013	−0.016
Gamma_1	0.015	−0.047	−0.009	−0.067	−0.003
Gamma_2	0.017	−0.018	−0.008	−0.017	0.084
Gamma_3	0.016	−0.031	0.008	0.047	0.105
Gamma_4	0.016	−0.037	0.001	−0.013	0.076
Gamma_5	0.018	−0.021	−0.012	−0.063	−0.014
Gamma_6	0.014	0.043	0.004	0.031	0.164
Gamma_7	0.018	−0.010	−0.001	0.003	0.053
Gamma_8	0.015	−0.039	−0.003	−0.003	0.083
Gamma_9	0.011	0.051	0.020	0.006	0.104
Gamma_10	−0.001	0.002	0.083	−0.104	0.208
Gamma_11	0.016	0.034	0.000	0.027	0.139
Gamma_12	0.011	0.056	−0.005	−0.005	0.130
Gamma_13	0.016	−0.018	0.019	0.121	0.056
Gamma_14	0.013	0.056	−0.011	−0.028	0.036
Gamma_15	0.017	−0.028	−0.009	−0.065	0.010
Gamma_16	0.018	−0.028	−0.001	−0.014	0.023
Gamma_17	0.017	−0.017	0.018	0.088	0.054
Gamma_18	0.017	−0.006	−0.001	0.005	0.053
Gamma_19	0.002	−0.002	0.195	0.031	−0.128
Gamma_20	0.016	−0.015	0.021	0.115	0.040

CHAPTER 6

Cognitive Intelligent Interactive Human-Machine Collaborative Situation Assessment and Threat Assessment Model

THIS CHAPTER ESTABLISHES A task scenario for cognitively intelligent, interactive, cooperative decision-making between manned aircraft and UAVs. Based on the definition of cooperative situation assessment and threat assessment, along with an analysis of the model structures, respective models are built and validated through case simulations.

6.1 SCENARIO OF HUMAN-MACHINE COLLABORATIVE DECISION-MAKING TASKS BASED ON COGNITIVE INTELLIGENT INTERACTION

Assumption: The UAV can make autonomous decisions based on the cognitive workload of the manned aircraft pilot, the external environment,

DOI: 10.1201/9781003791713-6

and its own status information. When necessary, the UAV can actively request decision-making assistance from the manned aircraft pilot, or the pilot can proactively engage in cooperative decision-making to jointly complete the combat mission.

Task Scenario: A single manned aircraft and four UAVs form a hybrid formation to execute a combat mission [91]. During the mission operation, the manned aircraft and UAVs need to cooperate in decision-making, such as situation assessment, threat assessment, task allocation, and route planning, as shown in Figure 6.1.

The area contains enemy targets. The overall mission execution process for the manned aircraft and UAVs hybrid formation is as follows:

(1) The manned aircraft pilot issues task commands and target locations to the available UAVs on standby at the base.

(2) Based on the target locations and known threat information, the UAVs perform route planning.

(3) The UAVs fly toward the target areas according to the predefined routes, accompanied by the manned aircraft.

(4) Upon reaching the target areas, the UAVs execute strike missions and report relevant information to the manned aircraft.

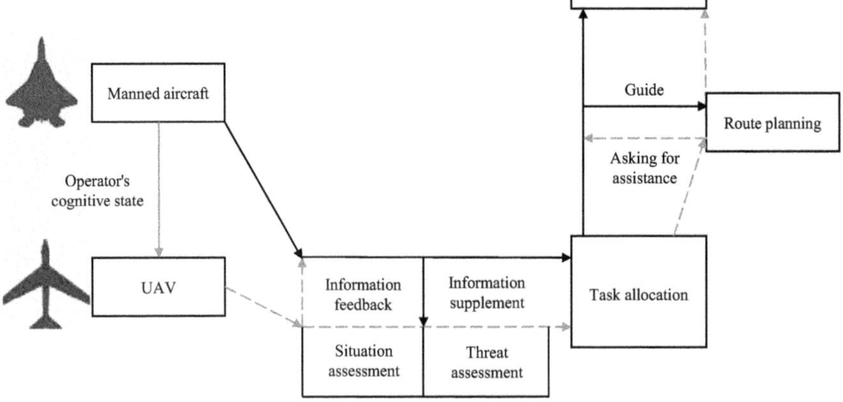

FIGURE 6.1 A scenario for cooperative decision-making between manned aircraft and UAVs

(5) After mission completion, the manned aircraft and UAVs return to the base along the planned routes and land.

(6) Throughout the mission operation, the manned aircraft and UAVs need to complete cooperative decision-making tasks such as situation assessment, threat assessment, task allocation, and route planning.

6.2 HUMAN-MACHINE COLLABORATIVE SITUATION ASSESSMENT MODEL BASED ON COGNITIVE INTELLIGENT INTERACTION

6.2.1 Definition and Model Structure of Collaborative Situation Assessment

Situation assessment belongs to decision-level information fusion, primarily providing a decision basis and support for command and control [92–94]. Regarding the definition of situation assessment, the information fusion model proposed by the Joint Directors of Laboratories (JDL) is widely accepted, as shown in Figure 6.2.

Situation Assessment constitutes the second level of the JDL model, facilitating deeper interpretation and prediction beyond simple target extraction. It clarifies critical information regarding the battlefield disposition of our and enemy forces, analyzes the operational posture of both sides, determines the distribution of hostile target groups, and anticipates enemy intentions.

In cognitive, intelligent, interactive, cooperative decision-making between manned aircraft and UAVs, the cognitive load of the manned aircraft pilot plays a crucial role in human-machine cooperative situation assessment. For instance, when the pilot's cognitive load is at a relatively

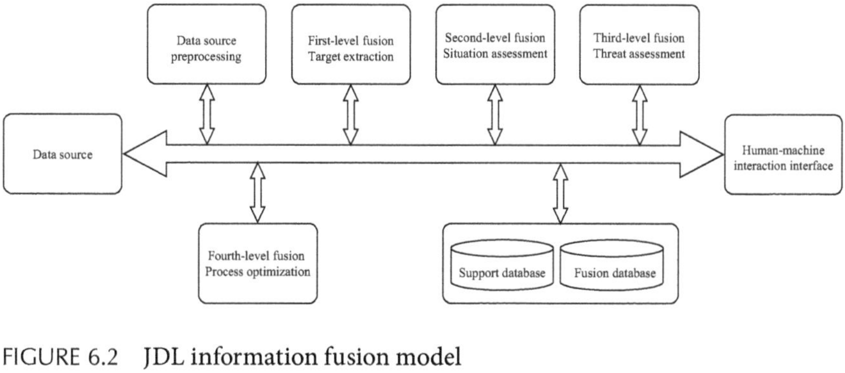

FIGURE 6.2 JDL information fusion model

low level, they can respond to sudden, urgent situations in human-machine cooperation more efficiently and maintain situational advantage with greater competence. Therefore, it is necessary to incorporate the manned aircraft pilot's cognitive workload information into the model of human-machine cooperative situation assessment.

In a cognitive-intelligent, interactive human-machine cooperative situation assessment system, UAVs conduct a comprehensive analysis of the current battlefield situation. This analysis incorporates factors such as weaponry and force characteristics of both sides, along with climate, terrain, and the cognitive load of the manned aircraft pilot, while also evaluating the interrelationships among these variables. The assessment results are fed back to the pilot, who then collaborates with the UAVs to complete the cooperative situation assessment, combining this with their own SA. The model structure is shown in Figure 6.3. Given the broad scope of situation assessment modeling, this chapter focuses on evaluating our force operational advantage as the specific research objective while also incorporating the pilot's cognitive load state as a key influencing factor.

6.2.2 Modeling Method for Cognitive Intelligent Interactive Cooperative Situation Assessment Based on IFCM

Classical Fuzzy Cognitive Maps (FCM) struggle to represent ambiguous relationships and possess limited modeling capacity. For instance, an event such as an enemy fighter intending to strike a friendly unit

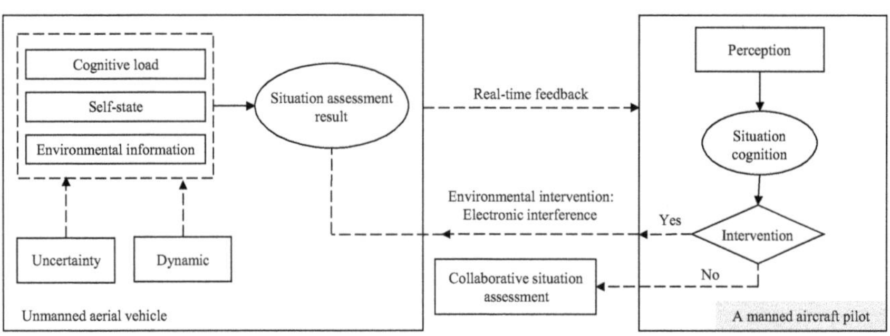

FIGURE 6.3 Model structure for cognitive intelligent interactive collaborative situation assessment

involves degrees of support, opposition, and uncertainty regarding its actual occurrence. Intuitionistic fuzzy sets can describe three aspects of information, including membership degree, nonmembership degree, and hesitancy degree. Furthermore, the intuitionistic fuzzy Ordered Weighted Averaging (OWA) operator can replace the general FCM threshold function to simulate the "AND/OR" relationships among fuzzy cognitive relations. Therefore, intuitionistic fuzzy sets are introduced to extend FCM, resulting in Intuitionistic Fuzzy Cognitive Maps (IFCM) to improve the adaptability of FCM modeling.

6.2.2.1 Concept of Intuitionistic Fuzzy Set

Let X be a nonempty set; then$A=\{\langle x,\mu_A(x),v_A(x)|x\in X\rangle\}$ is called an intuitionistic fuzzy set. $\mu_A(x)\in[0,1]$ and $v_A(x)\in[0,1]$represent the membership and nonmembership degrees of x in A, which is represented by the tuple $\langle\mu_A(x),v_A(x)\rangle$, satisfying $0\leq\mu_A(x)+v_A(x)\leq1$. Additionally, $\pi_A=1-\mu_A(x)-v_A(x)$ represents the hesitancy degree of element x in X toward A.

An example intuitionistic fuzzy number $\langle\mu_A(x),v_A(x)\rangle$, such as $\langle\mu_A(x),v_A(x)\rangle=\langle0.5,03\rangle$, indicates $\mu_A(x)=0.5$ and $v_A(x)=0.3$. This can be interpreted as "in a vote of 10 people on a certain event, 5 approve, 3 oppose, and the remaining 2 abstain".

6.2.2.2 Comparison of Intuitionistic Fuzzy Numbers

Suppose there are intuitionistic fuzzy numbers $a_1=\langle\mu_{a_1}(x),v_{a_1}(x)\rangle$ and $a_2=\langle\mu_{a_2}(x),v_{a_2}(x)\rangle$. Let $s(a_1)=\mu_{a_1}-v_{a_1}$ represent the score of a_1, and $h(a_1)=\mu_{a_1}+v_{a_1}$ represent the accuracy of a_1. The size relationship between them is as follows:

(1) If $s(a_1)>s(a_2)$, then $a_1>a_2$.

(2) If $s(a_1)=s(a_2)$, then:

If $h(a_1)=h(a_2)$, then $a_1=a_2$.

If $h(a_1)>h(a_2)$, then $a_1>a_2$.

If $h(a_1)<h(a_2)$, then $a_1<a_2$.

6.2.2.3 Operational Rules of Intuitionistic Fuzzy Numbers

Suppose intuitionistic fuzzy sets A and B exist on a nonempty set X. Then:

$$A \oplus B = \langle x, \mu_A(x) + \mu_B(x) - \mu_A(x)\mu_B(x), v_A(x)v_B(x) \rangle \tag{6.1}$$

$$A \otimes B = \langle x, \mu_A(x)\mu_B(x), v_A(x) + v_B(x) - v_A(x)v_B(x) \rangle \tag{6.2}$$

6.2.2.4 Ordered Weighted Averaging Operator

Suppose a set of intuitionistic fuzzy numbers is $a_i = \langle \mu_{a_i}(x), v_{a_i}(x) \rangle (i = 1, 2, \cdots, n)$. Then $\text{OWA}_e(a_1, a_2, a_3, \cdots, a_n) = \sum_{j=1}^{n} e_j b_j$, where $\text{OWA}_e(a_1, a_2, \cdots, a_n)$ is called the OWA operator. b_j is the j-th largest element in array $a_i (i = 1, 2, \cdots, n)$. $e = (e_1, e_2, \cdots, e_n)^T$ is the weighting vector associated with the OWA function, $e_j \in [0,1] (j = 1, 2, \cdots, n)$, and $\sum_{j=1}^{n} e_j = 1$.

Using intuitionistic fuzzy sets, the concept nodes and the adjacency weight matrix of the classical FCM carry out intuitive fuzzy processing, and the OWA operator replaces the general threshold function in the FCM model, receiving the IFCM model through $\text{IFCM} = (C, w, \text{OWA}_e)$.

Here, $C = (C_1, C_2, \cdots, C_n)$ represents the concept node of the IFCM, denoting the state space set. $C_i = \langle \mu_{oi}, v_{oi} \rangle$ represents the node state information. w is the fuzzy correlation matrix of concept nodes, and $w_{ij} = \langle \mu_{w_{ij}}, v_{w_{ij}} \rangle$ is the effect of C_j fromC_i. $\text{OWA}_e = (\text{OWA}_{e_1}, \text{OWA}_{e_2}, \cdots, \text{OWA}_{e_n})$ is an ordered weighted average operator, and the nodes $C_i (i = 1, 2, \cdots, n)$ correspond with OWA_{e_i}.

The initial state space of the IFCM is determined by the initial conditions. Through iterative computation, the model completes reasoning when the output state reaches a fixed value or a limited cycle state. The IFCM reasoning process is as follows:

Let the effect of node C_j on C_i at time t be r_{ij}.

When$i \neq j$:

$$r_{ij}(t) = \langle \mu_{r_{ij}(t)}, v_{r_{ij}(t)} \rangle = C_i(t) \otimes w_{ij}(t).$$

When $i = j$:

$$r_{ij}(t)\, r_{ij}(t) = C_j(t).$$

The value of node C_j at time $t+1$ is as follows:

$$C_j(t+1) = \text{OWA}_e\left[r_{1j}(t), r_{2j}(t), \cdots, r_{nj}(t)\right] = \sum_{k=1}^{n} e_k r_{\sigma(k)j}(t).$$

Here, $r_{\sigma(k)j}$ is the k-th intuitionistic fuzzy number with the descending order processing of $[r_{1j(t)}, r_{2j(t)}, \cdots, r_{nj}(t)]$, . . .

The application of IFCM primarily addresses uncertainty in battlefield environment information and the fuzziness of relationships between information. However, to incorporate model dynamics, Dynamic Fuzzy Cognitive Maps (DFCM) are introduced. Combining DFCM with IFCM forms the Dynamic Intuitionistic Fuzzy Cognitive Map (DIFCM).

The DIFCM reasoning process is as follows:

(1) Analyze and extract the model's concept nodes $C = (C_1, C_2, \cdots, C_n)$ as well as the environmental variables V.

(2) Determine the causal relationships between nodes (i.e., the adjacency weight matrix of $w = (w_{ij})_{n \times r}$).

(3) Determine the initial values of $C = (C_1, C_2, \cdots, C_n)$ and V, and determine the OWA operator and the related weight vector for the concept nodes.

(4) Perform reasoning until the output nodes reach a fixed value or a limited cycle state.

6.2.3 Assessment Model of Cognitive Intelligent Interactive Cooperative Situation

The cognitive workload of the manned aircraft pilot is introduced as a concept node into the situation assessment model. The DIFCM is employed to address uncertainty in information and the fuzziness of relationships within the situation assessment, establishing a cognitively intelligent, interactive human-machine cooperative situation assessment model. The manned aircraft pilot can intervene in the battlefield situation through control variables V to assist the UAV in understanding the current situation. The model is shown in Figure 6.4. The concept nodes and environmental control variables are listed in Table 6.1 and Table 6.2, respectively.

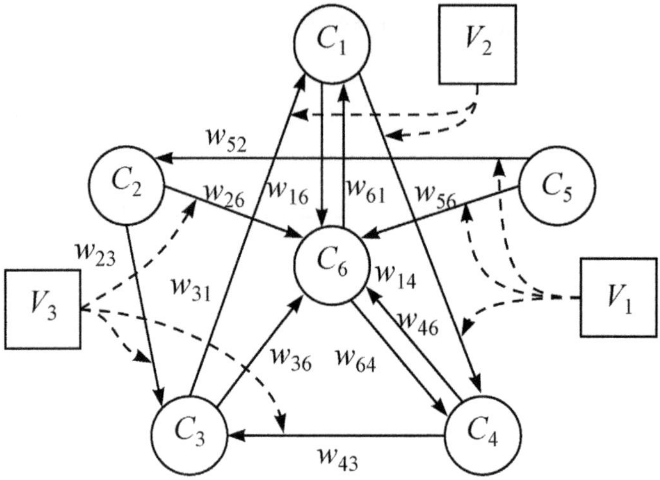

FIGURE 6.4 Cooperative situation assessment model based on DIFCM

TABLE 6.1 DIFCM Model Concept Nodes for Cooperative Situation Assessment

Node	Meaning
C_1	Enemy weapon system damage capability
C_2	Manned aircraft pilot's cognitive load
C_3	UAV weapon system
C_4	Manned aircraft and UAV rear support system
C_5	Manned aircraft and UAV information system
C_6	Our force situation advantage level

TABLE 6.2 Environmental Control Variables in the Cooperative Situation Assessment Model

Node	Meaning
V_1	Atmospheric density
V_2	Enemy support
V_3	Manned aircraft pilot intervention

The definition and calculation of the DIFCM model concept nodes and environmental control variables for cooperative situation assessment are as follows:

(1) Enemy Weapon Damage Capability

$$\begin{cases} \mu_{c_1} = \alpha \cdot \left(1 - \sum_{i=1}^{n}\left(\left(\frac{1}{2}\right)^{R_1^2/R_2^2}\right)^{m_i}\right) \\ \nu_{c_1} = \beta \cdot \left(\sum_{i=1}^{n}\left(\left(\frac{1}{2}\right)^{R_1^2/R_2^2}\right)^{m_i}\right) \end{cases} \tag{6.3}$$

where μ_{c_1} and ν_{c_1} are intuitionistic fuzzy set elements. α and β are correlation coefficients. R_1 is the kill radius of the enemy air defense weapon. R_2 is the hit accuracy of the enemy air defense weapon. m_i is the number of launches of the i-th enemy weapon type. n is the number of enemy weapon types.

(2) Manned Aircraft Pilot's Cognitive Load
This can be determined by the SVM-based manned aircraft pilot cognitive load assessment model from Chapter 5.

(3) UAV Weapon System
Assume the launch zone of the UAV's air-to-ground missile is as shown in Figure 6.5. R_{m1} is the minimum launch distance, R_{m2} is the maximum launch distance, and β_m is the maximum launch deviation angle.

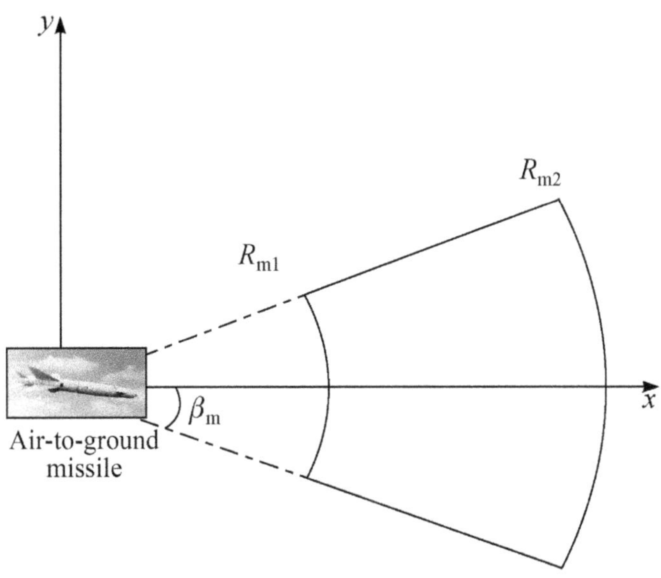

FIGURE 6.5 Schematic diagram of air-to-ground missile launch zone

Missile launch is simplified into two conditions:

(1) Must satisfy the required deviation angle: $\beta_t < \beta_m (\beta_t$ is the actual launch deviation angle).

(2) Must satisfy the launch boundary: $R_{m1} \le R \le R_{m2}$ (R is the actual launch distance).

After meeting these conditions, the kill radius and hit accuracy are considered as follows:

$$\begin{cases} \mu_{c_3} = \alpha \cdot \left(1 - \left(\left(\dfrac{1}{2}\right)^{R_3^2/R_4^2}\right)^{m_i}\right) \\ \nu_{c_3} = \beta \cdot \left(\left(\dfrac{1}{2}\right)^{R_3^2/R_4^2}\right)^{m_i} \end{cases} \tag{6.4}$$

Here, R_3 is the kill radius of the air-to-ground missile, R_4 is the hit accuracy, and m_i is the number of launches of the i-th type of air-to-ground missile.

(3) Manned Aircraft and UAV Rear Support System

The degree of support system relationship is fuzzified as shown in Table 6.3.

(4) Manned and UAV Information System

A simplification of the manned aircraft and UAVs information system is made. It suggests that the better the communication quality

TABLE 6.3 The Intuitionistic Fuzzy Set of the Degree of Aid System Relationship

Relationship Degree	Intuitionistic Fuzzy Set
Very high	⟨0.95,0.05⟩
High	⟨0.70,0.25⟩
Medium	⟨0.50,0.40⟩
Low	⟨0.25,0.70⟩
Very low	⟨0.05,0.95⟩
None	⟨0,1⟩

between manned aircraft and UAVs, the more reliable and stable the information system is. The calculation formula for the manned aircraft and UAVs information system is as follows:

$$\begin{cases} \mu_{c_5} = \alpha \cdot \tanh\left(\begin{array}{l} 0.5 \times \left(Q_{\max} - Q_{kx}\right) \times \left(t_f - t_p\right) + \\ 0.5 \times \dfrac{1}{n_r} \times \dfrac{\sum_{y=1}^{n_r} (M_r)_y}{\sum_{y=1}^{n_r} (M_t)_x} \end{array} \right) \\ v_{c_5} = \beta \cdot \left(1 - \tanh\left(\begin{array}{l} 0.5 \times \left(Q_{\max} - Q_{kx}\right) \times \left(t_f - t_p\right) + \\ 0.5 \times \dfrac{1}{n_r} \times \dfrac{\sum_{y=1}^{n_r} (M_r)_y}{\sum_{y=1}^{n_r} (M_t)_x} \end{array} \right) \right) \end{cases} \tag{6.5}$$

Here, $Q_{\max}$ is the maximum allowed data volume of the communication system, Q_{kx} is the lost data rate, t_f is the duration of a unit reception time slot, t_p is the propagation time per unit information, n_r is the number received, n_t is the number sent, M_t is the size of sent information, and M_r is the size of received information.

(5) Our Force Situation Advantage Degree

$$\text{The degree of our situation} = \begin{cases} \text{Special advantages}, & C_6 \geq [0.70 \sim 0.25] \\ \text{Advantages}, & [0.50, 0.40] \leq C_6 < [0.70, 0.25] \\ \text{Equilibrium}, & [0.40, 0.50] \leq C_6 < [0.50, 0.40] \\ \text{Disadvantage}, & [0.25, 0.70] \leq C_6 < [0.40, 0.50] \\ \text{Special disadvantage}, & [0, 1] \leq C_6 < [0.25, 0.70]. \end{cases} \tag{6.6}$$

In the DIFCM-based model, the environmental control variables V reflect the model's dynamics, encompassing both the influence of external environmental factors and the intervention by the manned aircraft operator.

6.2.3.1 Atmospheric Density

When the atmospheric density ρ is less than standard, the aircraft thrust decreases, ultimately affecting the maximum level flight speed and climb rate. The calculation formula for the variable of atmospheric density control is as follows:

$$V_1 = \begin{cases} 1 & \rho > 1.225 \\ \dfrac{\rho - 0.104}{1.225 - 0.104} & \rho < 1.225 \end{cases} \tag{6.7}$$

6.2.3.2 Enemy Support

Uses the distance between enemy support platforms and enemy targets as the main parameter. The calculation formula for designing the control variables V_2 of enemy support is as follows:

$$V_2 = \begin{cases} 0.7, \ d \in (d_1, d_2] \\ 0.4, \ d \in (d_2, d_3] \\ 0.1, \ d \in (d_3, d_4] \end{cases} \tag{6.8}$$

6.2.3.3 Manned Aircraft Pilot Intervention M_t

The control variable V_3 corresponds to intuitionistic fuzzy sets based on the intensity of operator intervention, as shown in Table 6.4.

The adjacency weight matrix in the situation assessment model is typically determined by domain experts. The DIFCM modeling method effectively captures the subjective hesitation characteristics of experts when determining the weights.

6.2.4 Case Simulation Analysis

Assume that the adjacency weight matrix for the situation assessment model is as follows:

TABLE 6.4 Intuitionistic Fuzzification of the Operator Intervention Degree

Intervention Degree	**Intuitionistic Fuzzy Set**
Very high	$\langle 0.95, 0.05 \rangle$
High	$\langle 0.70, 0.25 \rangle$
Medium	$\langle 0.50, 0.40 \rangle$
Low	$\langle 0.25, 0.70 \rangle$
Very low	$\langle 0.05, 0.95 \rangle$
None	$\langle 0, 1 \rangle$

$$w(\mu_{ij})=\begin{bmatrix}1&0&0&0.3&0&0.65\\0&1&0.6&0&0&0.7\\0.5&0&1&0&0&0.7\\0&0&0.45&1&0&0.8\\0&0.63&0&0&1&0.75\\0.3&0&0&0.2&0&1\end{bmatrix}\tag{6.9}$$

$$w(v_{ij})=\begin{bmatrix}0&1&1&0.6&1&0.25\\1&0&0.2&1&1&0.1\\0.4&1&0&1&1&0.25\\1&1&0.35&0&1&0.15\\1&0.28&1&1&0&0.15\\0.5&1&1&0.7&1&0\end{bmatrix}\tag{6.10}$$

Influence functions and parameters for different environmental variables V_k on the corresponding weights w_{ij} are designed, as shown in Table 6.5.

In Table 6.5, environmental control variables V_1 and V_2 simultaneously affect w_{14}. In such cases, the variable with the greatest influence on w_{14} is selected as the influencing variable.

6.2.4.1 Model Dynamics Verification

Environmental control variables at times t_1 and t_2 are shown in Table 6.6.

TABLE 6.5 Influence Functions of w_{ij} in the Collaborative Situation Assessment DIFCM V_k

w_{ij}	The Influence of V_k on w_{ij}	Influence Function
w_{14}	$f(w_{14},V_1,0.2)$ $f(w_{14},V_2,0.33)$	$w_{14}-0.2\cdot V_1\cdot w_{14}$ $w_{14}+0.33\cdot V_2\cdot w_{14}$
w_{23}	$f(w_{23},V_3,0.3)$	$w_{23}+0.3\cdot V_3\cdot w_{23}$
w_{26}	$f(w_{26},V_3,0.4)$	$w_{26}+0.4\cdot V_3\cdot w_{26}$
w_{31}	$f(w_{31},V_1,0.2)$	$w_{31}-0.2\cdot V_2\cdot w_{31}$
w_{43}	$f(w_{43},V_3,0.5)$	$w_{43}+0.5\cdot V_3\cdot w_{43}$
w_{52}	$f(w_{52},V_1,0.25)$	$w_{52}-0.25\cdot V_1\cdot w_{52}$
w_{56}	$f(w_{56},V_1,0.15)$	$w_{56}-0.15\cdot V_1\cdot w_{56}$

TABLE 6.6 Values of V_k at Aimes t_1 and t_2

V_k	t_1	t_2
Atmospheric density V_1	0.1	0.4
Enemy reinforcements V_2	0.2	0.6
Operator intervention V_3	0.5	0.1

TABLE 6.7 DIFCM Model Weights at Times t_1 and t_2

w_{ij}	w_{ij}^0	$w_{ij}^{t_1}$	$w_{ij}^{t_2}$
w_{14}	⟨0.3,0.6⟩	⟨0.3066,0.6120⟩ ⟨0.2802,0.5604⟩	⟨0.3240,0.6480⟩ ⟨0.2406,0.4812⟩
w_{23}	⟨0.6,0.2⟩	⟨0.6360,0.2120⟩	⟨0.7080,0.2360⟩
w_{26}	⟨0.7,0.1⟩	⟨0.8400,0.1200⟩	⟨0.7280,0.1040⟩
w_{31}	⟨0.5,0.4⟩	⟨0.4900,0.3200⟩	⟨0.4700,0.3760⟩
w_{43}	⟨0.45,0.35⟩	⟨0.5625,0.4375⟩	⟨0.4725,0.3675⟩
w_{52}	⟨0.63,0.28⟩	⟨0.6458,0.2870⟩	⟨0.6930,0.3080⟩
w_{56}	⟨0.68,0.3⟩	⟨0.7613,0.1523⟩	⟨0.7950,0.1590⟩

Analysis shows that the external environment at time t_1 is superior to that at t_2. Given the model's initial input state values, cooperative situation assessment results for times t_1 and t_2 can be obtained. DIFCM model weights at times t_1 and t_2 are shown in Table 6.7.

Iterative reasoning according to the DIFCM model yields the results shown in Table 6.8. The first set represents the initial input node state values. The second set represents the model reasoning result without considering environmental effects. The third and fourth sets represent the model reasoning results considering the environment at times t_1 and t_2, respectively. Compared to the second set, the membership and nonmembership degrees of the friendly situation advantage degree C_6 in the third and fourth sets are improved, indicating a greater friendly advantage at time t_1. Comparing the model outputs at t_1 and t_2, the changes in each concept node state value are shown in Figure 6.6. It can be seen that the external environment at time t_2 is characterized by greater enemy support and less operator intervention. The model reasoning results show that the state values of nodes C_2 andC_3 at time t_2 are larger than at time t_1. Thus, the simulation results are consistent with the qualitative analysis.

TABLE 6.8 Concept Node Values at Times t_1 and t_2 Considering Environmental Influence

		Model Inference Results		
Node	**Initial Value**	**No Environment**	t_1	t_2
C_1	<0.55,0.45>	<0.4500,0.4412>	<0.4488,0.4401>	<0.4465,0.4379>
C_2	<0.75,0.25>	<0.7063,0.2800>	<0.7075,0.2856>	<0.7111,0.2874>
C_3	<0.7, 0.3>	<0.6506,0.2364>	<0.6553,0.2396>	<0.6566,0.2385>
C_4	<0.65,0.35>	<0.4869,0.3448>	<0.4871,0.3452>	<0.4878,0.3462>
C_5	<0.7,0.3>	<0.7000,0.2800>	<0.7000,0.2800>	<0.7000,0.2800>
C_6	<0.5,0.5>	<0.4916,0.3788>	<0.5652,0.3902>	<0.5358,0.4039>

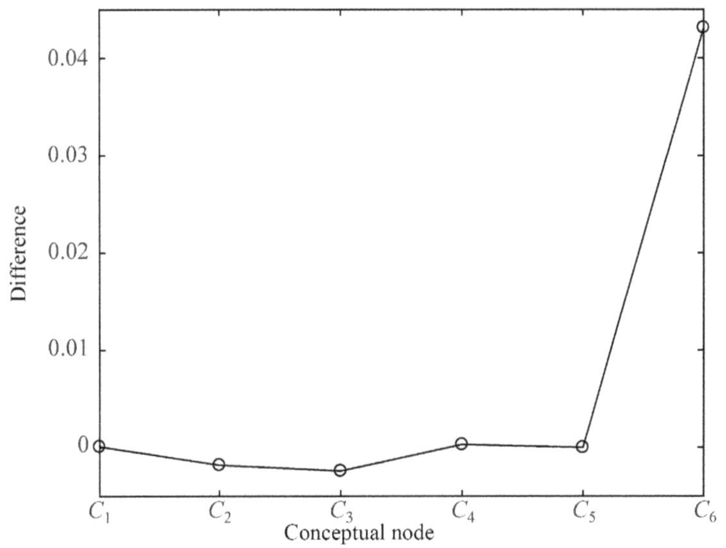

FIGURE 6.6 Comparison of concept node values at times t_1 and t_2

6.2.4.2 Model Validity Verification about Uncertainty Reasoning Ability

The initial input node state values are set as shown in Table 6.9. In this case, the state values of the first group of concept nodes do not have a hesitation degree. The second group of concept nodes C_1 has a hesitancy degree of 0.1. The third group then sets a hesitancy degree of 0.2 to C_2. The fourth group also sets a hesitancy degree of 0.2 to C_3. The fifth group sets a hesitancy degree of 0.3 to C_4. It suggests that, from the first group to the fifth group, the number of concept nodes with hesitancy degrees in the

TABLE 6.9 Presents Different Model Inputs and Outputs with Varying Degrees of Uncertainty

Node	1	2	3	4	5
C_1	<0.65,0.35>	<0.7,0.2>	<0.7,0.2>	<0.7,0.2>	<0.7,0.2>
C_2	<0.7,0.3>	<0.7,0.3>	<0.6,0.2>	<0.6,0.2>	<0.7,0.3>
C_3	<0.7, 0.3>	<0.7,0.3>	<0.7,0.3>	<0.7,0.1>	<0.7, 0.3>
C_4	<0.6,0.4>	<0.6,0.4>	<0.6,0.4>	<0.6,0.4>	<0.6,0.1>
C_5	<0.7,0.3>	<0.7,0.3>	<0.6,0.2>	<0.6,0.2>	<0.7,0.3>
C_6	<0.5380, 0.4293>	<0.5332, 0.4216>	<0.4857, 0.3674>	<0.4838, 0.3549>	<0.4479, 0.3026>

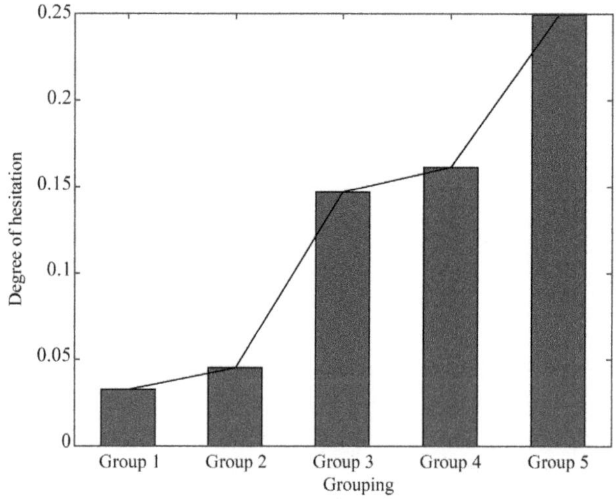

FIGURE 6.7 Comparison of hesitancy degrees of node output for each group

model's reasoning output is constantly increasing. The hesitancy degrees of the output for each group of nodes C_6 are shown in Figure 6.7.

6.2.4.3 Verification of the Inference Results of the Situation Assessment Model

The result of the situation awareness model reasoning is an intuitive fuzzy number. The degrees of confidence and accuracy are used to distinguish the situation assessment results. The input and output values of the model reasoning are shown in Table 6.10.

The first group was set as the control group, while the second group had stronger enemy weapon damage capability. The third group had a higher

TABLE 6.10 Input and Output of Different Situations

Node	1	2	3	4
C_1	<0.65,0.35>	<0.9,0.1>	<0.65,0.35>	<0.65,0.35>
C_2	<0.7,0.2>	<0.7,0.2>	<0.6,0.3>	<0.8,0.1>
C_3	<0.6,0.3>	<0.6,0.3>	<0.5,0.4>	<0.8,0.1>
C_4	<0.7,0.2>	<0.7,0.2>	<0.7,0.2>	<0.7,0.2>
C_5	<0.6,0.3>	<0.6,0.3>	<0.6,0.3>	<0.75,0.12>
C_6	<0.5294,0.3573>	<0.5120,0.3580>	<0.4936,0.3933>	<0.6080,0.2596>

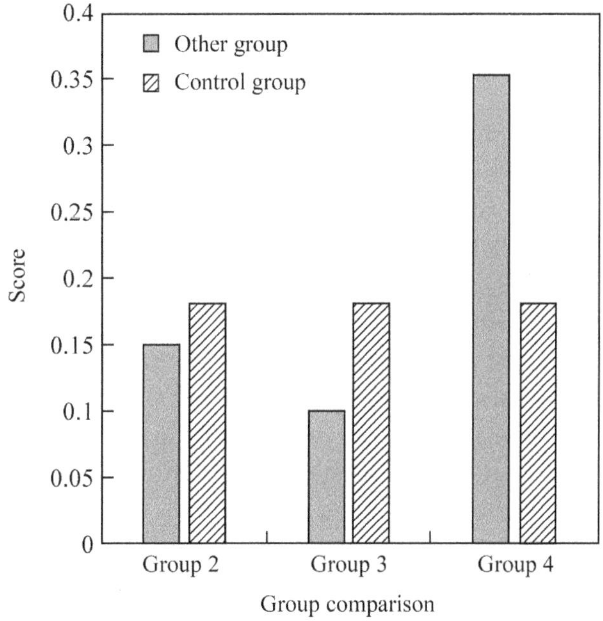

FIGURE 6.8 The scores of the group model are compared with those of the control group

cognitive workload for our aircraft operators and a weaker weapon system, and the fourth group had a lower cognitive workload for our aircraft operators and a stronger weapon system. The scores of the model reasoning output for each group are shown in Figure 6.8.

It suggests that the scores of the second group are lower than those of the control group, indicating that our situation is slightly inferior to the enemy's. The scores of the third group are even lower than those of the control group, suggesting that our situation is in a disadvantageous position. The scores of the fourth group are much higher than those of the control group, indicating that our situation has a significant advantage. These are consistent with the qualitative analysis and judgment of the superiority or inferiority of the enemy and our situations, verifying the effectiveness of the collaborative situation awareness model.

6.3 HUMAN-MACHINE COOPERATIVE THREAT ASSESSMENT MODEL BASED ON COGNITIVE INTELLIGENT INTERACTION

Threat assessment is a higher level of information fusion, belonging to the third level of the JDL model. It involves assessing the threat level of enemy targets based on the results of situation assessment, as well as providing decision support for command and control.

6.3.1 Definition of Cooperative Threat Assessment and Model Structure

Threat assessment involves three steps, including the extraction of threat elements, the calculation of threat degree, and the determination of threat level, as shown in Figure 6.9.

Threat element extraction: For different battlefield tasks, the threat assessment elements to be extracted also vary greatly. For example, in the task of suppressing enemy air defense forces (SEAD), the main threat assessment elements include the type of enemy ground air defense weapons, the number of air defense weapons, the position of enemy radars, and the interference capability, to name a few [95].

Threat degree calculation: Calculate the extracted assessment elements and quantitatively describe the threat degree of enemy combat units against us. The quantified assessment elements can facilitate the establishment of a calculation model and better analyze the impact of each element on the threat level.

Threat level determination: Based on different threat sources and the weights set by expert experience, perform a quantitative calculation of the

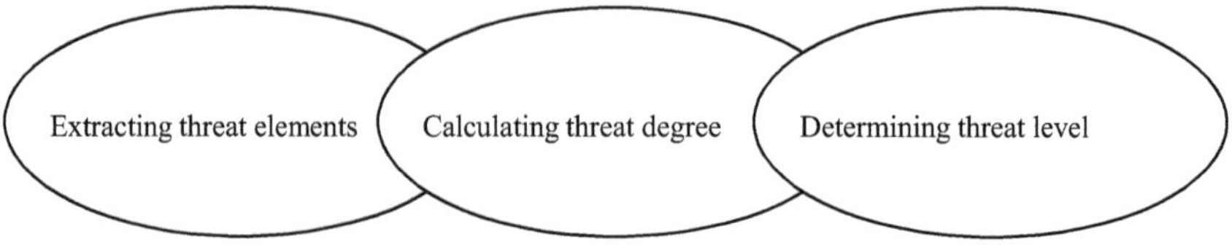

FIGURE 6.9 Threat assessment three-level function model

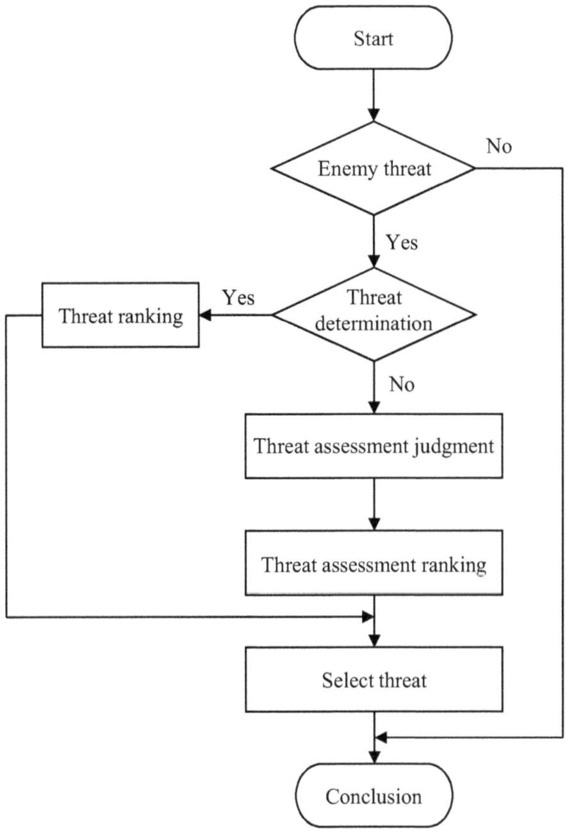

FIGURE 6.10 Threat assessment process diagram

threat degree, and determine the level and ranking of the threat degree. The specific steps of threat assessment are shown in Figure 6.10.

In the modeling of collaborative threat assessment between manned aircraft and unmanned aircraft, the following issues need to be addressed:

(1) It is necessary to have the ability to handle the uncertainty of complex battlefield environment information.

(2) It is necessary to utilize the correlation characteristics of data to improve the accuracy of the assessment.

(3) In the determination of the model's weights, both expert knowledge and data should be combined.

(4) The cognitive workload of the manned aircraft operator should be regarded as one of the elements of the assessment.

6.3.2 Cognitive Intelligent Interactive Collaborative Threat Assessment Model Structure

UAVs assess the threat level of enemy targets based on relevant target information and the pilot's cognitive load state. Under the cognitive intelligent interaction mode, UAVs dynamically perform threat assessment by integrating the cognitive load state of the manned aircraft pilot, thereby providing decision-making assistance. When necessary, the manned aircraft pilot can augment the UAV's threat assessment by integrating their own experience, support data from the manned platform, and information shared by other friendly combat units. The overall model structure is illustrated in Figure 6.11.

6.3.3 Threat Assessment Modeling Method Based on IFCM and Genetic Algorithm

To address the uncertainty issues in threat assessment, an Intuitionistic Fuzzy Cognitive Map (IFCM) is used for model construction. Simultaneously, the model is integrated with a genetic learning algorithm to reduce subjectivity in determining the adjacency weight matrix. This learning model can incorporate causal relationship weights between nodes in the IFCM. The main advantages include applicability in high-dimensional continuous spaces, the availability of richer evolutionary operators during the search process, and strong global optimization capabilities. The modeling flowchart combining IFCM and a genetic learning algorithm is shown in Table 6.11.

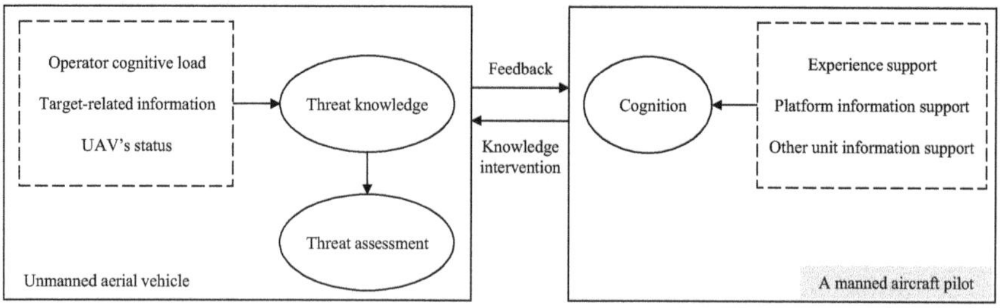

FIGURE 6.11 Structure of the cognitive intelligent interactive cooperative threat assessment model

TABLE 6.11 Steps for Building the Genetic Algorithm-Based IFCM Model

Step	Content
1	Input historical data, i.e., concept node state values ($C_1, C_2, ..., C_n$).
2	Generate random values for the adjacency matrix W.
3	Iteration: Perform genetic algorithm learning on the adjacency matrix W until convergence criteria are met or learning continues.
4	If the fitness function exceeds the maximum fitness or the iteration count exceeds the maximum, exit the iteration.
5	Return: Optimal weight matrix W

The implementation steps of the genetic algorithm-based cooperative threat assessment IFCM model are as follows:

(1) Select threat assessment elements as concept nodes of IFCM based on expert experience, determine the causal relationship between nodes, and obtain the initial adjacency weight matrix W.

(2) Design the chromosome for the genetic algorithm and establish the initial population. Since IFCM includes membership and nonmembership degrees, if there are m unknown weights and n nodes, the chromosome length is $2\times m+n$. If the population size is p, the initial population is $p\times(2\times m+n)$.

(3) Define the fitness function. The general form of the fitness function is as follows:

$$\text{fitness} = h(\text{Error}) \tag{6.11}$$

$$h(x) = \frac{1}{\alpha x + 1} \tag{6.12}$$

Here, α is a parameter controlling precision, and Error is the difference between the input and the output vectors.

$$\text{Error} = \frac{1}{N}\sum_{n=1}^{N}\left|C_n(t) - C_n(t+1)\right|^2 \tag{6.13}$$

In this case, N denotes the number of nodes in DIFCM. The smaller the error is, the closer the fitness function is to 1.

(4) Determine the selection strategy. The roulette wheel strategy is used here, where the probability of an individual being selected is related to its fitness function value. The higher the fitness is, the greater the probability of selection.

(5) Design the crossover operator. Methods such as single-point crossover, partially matched crossover, or cycle crossover can be used.

(6) Design the mutation operator.

(7) Design the iteration termination conditions. These are typically based on the fitness function value or the maximum number of iterations.

6.3.4 Cognitive Intelligent Interactive Cooperative Threat Assessment Model

Based on the selection of threat elements for our UAVs and incorporating the manned aircraft pilot's cognitive workload as a concept node, the threat assessment IFCM model is established, as demonstrated in Figure 6.12.

6.3.4.1 Positional Threat of Target to UAVs

Positional threat is generally considered from three factors, including the distance between the enemy and us, the enemy's speed, and the enemy's movement direction.

$$C_1 = \alpha \cdot E_e + \beta \cdot E_d + \gamma \cdot E_h \tag{6.14}$$

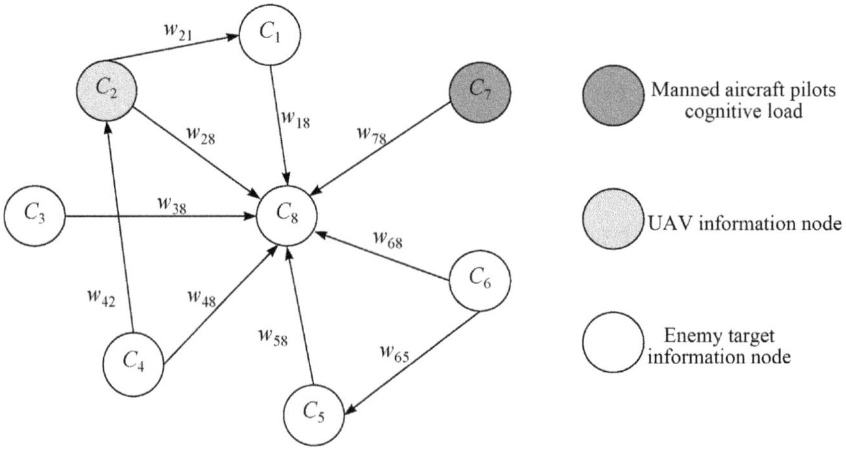

FIGURE 6.12 IFCM threat assessment model

Here, E_e, E_d, and E_h represent angle threat, distance threat, and velocity threat, respectively.α, β, and γ are corresponding coefficients.

6.3.4.1.1 Distance Threat Normalization Function

$$E_d = \begin{cases} 0.5 - 0.2 \times \dfrac{r - rmt}{rm - rmt} & rmt < r < rm \\ 0.5 & r \le rm, r < rmt \\ 0.8 & \max(rm, rmt) < r < rr \\ 1.0 & rm < r < rmt \end{cases} \tag{6.15}$$

Here, r is the distance to the enemy target, rm is the maximum range of the UAV's missile. rmt is the minimum range of the enemy's air defense missile. rr is the maximum tracking distance of the UAV's radar.

6.3.4.1.2 Angle Threat Calculation Formula

$$E_e = |\theta_R| / 2\pi \tag{6.16}$$

6.3.4.1.3 Velocity Threat Calculation Formula

$$E_v = \alpha \cdot \frac{v}{vm} \tag{6.17}$$

6.3.4.2 Maneuverability of the UAVs

$$C_2 = \varepsilon_c \cdot B \tag{6.18}$$

ε_c is an operational coefficient derived from the database. B is the mobility parameter.

$$B = 0.1 \cdot \overline{\omega} + 0.15 \cdot \overline{Ma} + 0.15 \cdot \overline{T_R} + 0.25 \cdot \overline{n_y} + 0.35 \cdot \overline{P_{SEP}} \tag{6.19}$$

$\overline{\omega}$ is the maximum instantaneous angular velocity of the UAV. $\overline{Ma}$ is the maximum Mach number. $\overline{T_R}$ is the thrust-to-weight ratio. $\overline{n_y}$ is the maximum sustained turn load factor. $\overline{P_{SEP}}$ is the maximum specific excess power. "–" denotes normalized parameters using the sigmoid function.

6.3.4.3 *Electronic Countermeasure Capability of the Target*

$$C_3 = K_x \cdot \left(\frac{\overline{W}}{\overline{Q}}\right) \cdot \left(\frac{\overline{F}}{\overline{f}}\right) \cdot \left(\frac{\overline{R}}{\overline{r}}\right) \cdot P_{yd} \tag{6.20}$$

K_x is the jamming factor. $\overline{Q}$, $\overline{f}$ and $\overline{r}$ are the actual spatial jamming range, frequency jamming range, and jamming exposure distance. $\overline{W}$, $\overline{F}$ and $\overline{R}$ are the required spatial and frequency jamming ranges, and the jamming exposure distance for the mission, respectively. P_{yd} is the guidance success rate, and "–" represents normalization.

6.3.4.4 *Type of the Target*

Assume there are two types of ground air defense threats: surface-to-air missiles (SAM) and anti-aircraft artillery (AAA). Five typical threat sources are defined here, as shown in Table 6.12.

The ground air defense weapon types (T_1, T_2, T_3, T_4, T_5) correspond one-to-one with radar types (R_1, R_2, R_3, R_4, R_5). UAVs can generally identify the target type by detecting the enemy radar type through radar reconnaissance. Figure 6.13 shows that radar type can be judged from four aspects.

TABLE 6.12 Information on Typical Air Defense Weapons

Threat Source	Horizontal Range/km	Vertical Range/km	Radar Type	Corresponding Air Defense System
T_1	23.4~27	21.6	R_1 (EM004)	SAM1/SAM2
T_2	14.4	19.8	R_2 (EM007)	SAM3
T_3	7.2	19.8	R_3 (EM008)	SAM4
T_4	3.6	6.0	R_4 (EM010)	AAA3
T_5	3.6	3.0	R_5	AAA1/AAA2

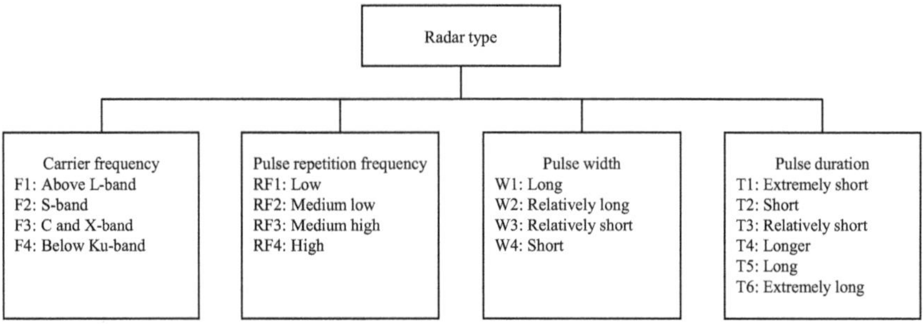

FIGURE 6.13 Radar identification parameters

TABLE 6.13 Fuzzy Threat Values for Target Types

Threat Source Type	Fuzzy Threat Value
T_1	0.9
T_2	0.7
T_3	0.5
T_4	0.3
T_5	0.1

The fuzzy threat values correspond to different target types and are determined in Table 6.13.

6.3.4.5 *Lethality of the Target's Weaponry*

Conditions for enemy air defense systems (missiles and artillery) to engage our UAVs include:

(1) Enemy warning radar detection capability exceeds a certain value, typically 0.6.

(2) The UAV is within the enemy's engageable range: $L_{min} < R < L_{max}$, where L_{max} and L_{min} are the maximum and minimum ranges of the enemy air defense weapon, respectively.

(3) The UAV is within the engageable altitude of the enemy air defense weapon: $H_{min} < h < H_{max}$, where H_{max} and H_{min} are the maximum and minimum altitudes of the enemy air defense weapon, respectively.

(4) The UAV is within the engageable angle of the enemy air defense weapon: $\alpha < \beta_{max}$, where β_{max} is the maximum angle of the air defense weapon.

When the enemy air defense weapon meets the conditions to attack our UAV, its lethality is calculated as follows:

$$C_5 = 1 - \left(\left(\frac{1}{2} \right)^{R_1^2 / R_2^2} \right)^{m_0} \tag{6.21}$$

R_1 is the kill radius of the weapon and R_2 is the hit probability of the weapon.

6.3.4.6 Radar Detection Capability of the Target

Enemy ground target radar operation requires meeting the following conditions:

(1) Our UAV flies within the enemy radar's operational range: $R_{ur} < R_{work}$ (where R_{work} is the maximum operational distance of the enemy radar, and R_{ur} is the distance between the UAV and the enemy radar).

(2) The UAV enters the enemy radar's scanning area: $\alpha_t < \alpha_r$ and $\beta_t < \beta_r$ (where α_t and β_t are the angles between the UAV and the enemy radar in the vertical and horizontal planes, respectively. α_r and β_r are the maximum scanning angles of the enemy radar in the vertical and horizontal planes).

When both conditions are satisfied, the enemy radar's detection capability is calculated as follows:

$$C_6 = \exp\left[\frac{(R_{ur} / R_{work})^4 \ln P_{do} \ln P_{fa}}{\ln P_{fa} - \ln P_{do} + (R_{ur} / R_{work})^4 \ln P_{do}}\right] \tag{6.22}$$

Here, P_{fa} is the false alarm probability of the enemy radar. P_{do} is the detection probability at maximum range. If the conditions are not met, then $C_6 = 0$.

6.3.4.7 Cognitive Load of the Manned Aircraft Operator

It can be determined using the SVM-based manned aircraft operator cognitive load assessment model from Chapter 5.

6.3.5 Case Simulation Validation

Assume the initial adjacency weight matrix for the cooperative threat assessment model is as follows:

$$W = \begin{bmatrix} \langle 1,0\rangle & \langle 0,1\rangle & \langle 0,1\rangle & \langle 0,1\rangle & \langle 0,1\rangle & \langle 0,1\rangle & \langle 0,1\rangle & w_{18} \\ w_{21} & \langle 1,0\rangle & \langle 0,1\rangle & \langle 0,1\rangle & \langle 0,1\rangle & \langle 0,1\rangle & \langle 0,1\rangle & w_{28} \\ \langle 0,1\rangle & \langle 0,1\rangle & \langle 1,0\rangle & \langle 0,1\rangle & \langle 0,1\rangle & \langle 0,1\rangle & \langle 0,1\rangle & w_{38} \\ \langle 0,1\rangle & w_{42} & \langle 0,1\rangle & \langle 1,0\rangle & \langle 0,1\rangle & \langle 0,1\rangle & \langle 0,1\rangle & w_{48} \\ \langle 0,1\rangle & \langle 0,1\rangle & \langle 0,1\rangle & \langle 0,1\rangle & \langle 1,0\rangle & \langle 0,1\rangle & \langle 0,1\rangle & w_{58} \\ \langle 0,1\rangle & \langle 0,1\rangle & \langle 0,1\rangle & \langle 0,1\rangle & w_{65} & \langle 1,0\rangle & \langle 0,1\rangle & w_{68} \\ \langle 0,1\rangle & \langle 0,1\rangle & \langle 0,1\rangle & \langle 0,1\rangle & \langle 0,1\rangle & \langle 0,1\rangle & \langle 1,0\rangle & w_{78} \\ \langle 0,1\rangle & \langle 0,1\rangle & \langle 0,1\rangle & \langle 0,1\rangle & \langle 0,1\rangle & \langle 0,1\rangle & w_{87} & \langle 1,0\rangle \end{bmatrix} \tag{6.23}$$

TABLE 6.14 Value Ranges for Weights Determined by Experts

w_{18}	w_{21}	w_{28}	w_{38}
$\langle[0.6,0.7],[0.2,0.3]\rangle$	$\langle[0.2,0.4],[0.5,0.6]\rangle$	$\langle[0.5,0.6],[0.3,0.4]\rangle$	$\langle[0.7,0.8],[0.1,0.2]\rangle$
w_{42}	w_{48}	w_{58}	w_{65}
$\langle[0.3,0.4],[0.5,0.6]\rangle$	$\langle[0.6,0.7],[0.2,0.3]\rangle$	$\langle[0.7,0.8],[0.1,0.2]\rangle$	$\langle[0.5,0.7],[0.1,0.2]\rangle$
w_{68}	w_{78}	w_{87}	
$\langle[0.4,0.6],[0.3,0.4]\rangle$	$\langle[0.2,0.4],[0.5,0.6]\rangle$	$\langle[0.4,0.5],[0.3,0.5]\rangle$	

The value ranges for the weights to be determined by experts are shown in Table 6.14.

Here, 100 sets of concept node values were selected from the experimental historical database. Seventy sets were randomly chosen to train the IFCM adjacency matrix weights, and the remaining 30 sets were used to test the model's accuracy.

(1) Chromosome Encoding. There are 11 sets of weights in the model that need expert determination. The chromosome encoding is as follows:

$$\text{chromosome} = [\mu_{18}, \nu_{18}, \mu_{21}, \nu_{21}, \mu_{28}, \nu_{28}, \cdots, \mu_{87}, \nu_{87}] \tag{6.24}$$

μ and ν are the membership and nonmembership degrees of the corresponding i-th weight.

(2) Selection. As shown in Figure 6.14, the roulette wheel rule divides the wheel into segments proportional to chromosome fitness. Each spin selects a chromosome. Seventy sets form a population. The probability of an individual i being selected is as follows:

$$P_i = \frac{fitness(i)}{\sum_{i=1}^{P} fitness(i)} \tag{6.25}$$

(3) Crossover. For the newly generated population, crossover is performed by randomly selecting two chromosomes and performing a single-point crossover at a random position, as shown in Figure 6.15. Crossover is controlled by the crossover probability.

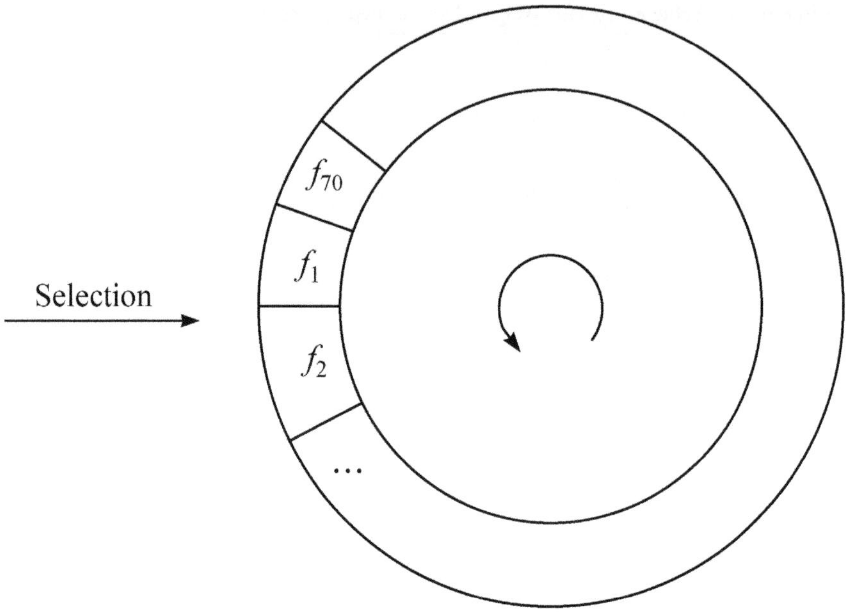

FIGURE 6.14 Schematic diagram of roulette wheel selection

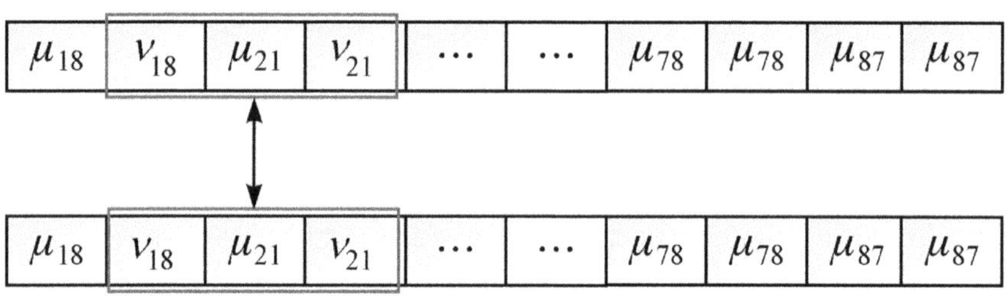

FIGURE 6.15 Schematic diagram of crossover

(4) Mutation. A random number is generated. If this number is less than the mutation probability, mutation is performed at a random position on the chromosome, as shown in Figure 6.16.

(5) Elitism Strategy. To preserve good genes, an elitism strategy is used, in which chromosomes with the best fitness are stored separately. After the selection, crossover, and mutation operations, these stored chromosomes replace the chromosomes with the worst fitness in the population.

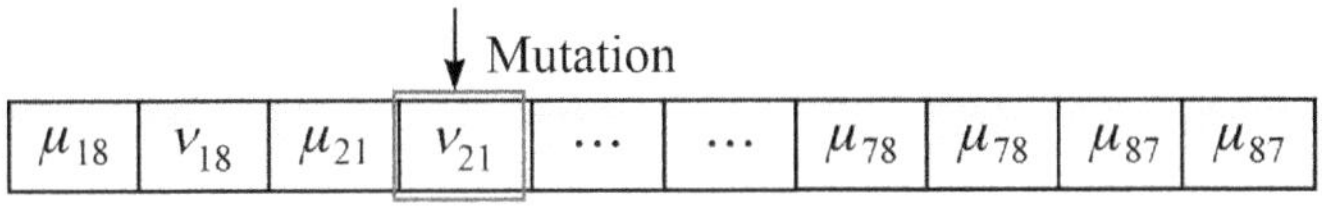

FIGURE 6.16 Schematic diagram of mutation

TABLE 6.15 Partial Training Data

Serial Number	C_1	C_2	C_3	C_4	C_5	C_6	C_7	C_8
1	0.1	0.41	0.7	0.5	0.6	0.67	0.5	0.4210
2	0.1	0.41	0.68	0.7	0.5	0.58	0.5	0.5356
3	0.1	0.53	0.75	0.4	0.3	0.5	0.5	0.3899
4	0.1	0.65	0.73	0.33	0.58	0.48	0.5	0.4496
5	0.1	0.75	0.49	0.57	0.78	0.8	0.5	0.5579
6	0.3	0.46	0.55	0.67	0.75	0.64	0.5	0.5678
7	0.3	0.58	0.35	0.45	0.21	0.35	0.5	0.2374
8	0.3	0.62	0.78	0.79	0.89	0.79	0.5	0.8125
9	0.3	0.37	0.59	0.38	0.41	0.62	0.5	0.3795
10	0.3	0.29	0.57	0.56	0.67	0.49	0.5	0.5312
11	0.5	0.59	0.48	0.78	0.45	0.67	0.5	0.6549
12	0.5	0.67	0.72	0.32	0.59	0.63	0.5	0.5984

(6) Check Iteration Termination. The iteration termination condition is set so that the error value Error for a chromosome is less than a threshold $Error < \varepsilon$, here$\varepsilon = 0.0006$. And the partial training data are shown in Table 6.15.

The final target chromosome obtained through the genetic algorithm yields the weight matrix:

$$W = \begin{bmatrix} \langle 1,0\rangle & \langle 0,1\rangle & \langle 0,1\rangle & \langle 0,1\rangle & \langle 0,1\rangle & \langle 0,1\rangle & \langle 0,1\rangle & \langle 0.6127,0.2135\rangle \\ \langle 0.3519,0.5287\rangle & \langle 1,0\rangle & \langle 0,1\rangle & \langle 0,1\rangle & \langle 0,1\rangle & \langle 0,1\rangle & \langle 0,1\rangle & \langle 0.5569,0.3124\rangle \\ \langle 0,1\rangle & \langle 0.37860.5403\rangle & \langle 1,0\rangle & \langle 0,1\rangle & \langle 0,1\rangle & \langle 0,1\rangle & \langle 0,1\rangle & \langle 0.7436,0.1105\rangle \\ \langle 0,1\rangle & \langle 0,1\rangle & \langle 0,1\rangle & \langle 1,0\rangle & \langle 0,1\rangle & \langle 0,1\rangle & \langle 0,1\rangle & \langle 0.6824,0.2637\rangle \\ \langle 0,1\rangle & \langle 0,1\rangle & \langle 0,1\rangle & \langle 0,1\rangle & \langle 1,0\rangle & \langle 0,1\rangle & \langle 0,1\rangle & \langle 0.7423,0.1226\rangle \\ \langle 0,1\rangle & \langle 0,1\rangle & \langle 0,1\rangle & \langle 0,1\rangle & \langle 0.6320,0.1150\rangle & \langle 1,0\rangle & \langle 0,1\rangle & \langle 0.4861,0.3796\rangle \\ \langle 0,1\rangle & \langle 0,1\rangle & \langle 0,1\rangle & \langle 0,1\rangle & \langle 0,1\rangle & \langle 0,1\rangle & \langle 1,0\rangle & \langle 0.3428,0.5319\rangle \\ \langle 0,1\rangle & \langle 0,1\rangle & \langle 0,1\rangle & \langle 0,1\rangle & \langle 0,1\rangle & \langle 0,1\rangle & \langle 0,1\rangle & \langle 1,0\rangle \end{bmatrix} \quad (6.26)$$

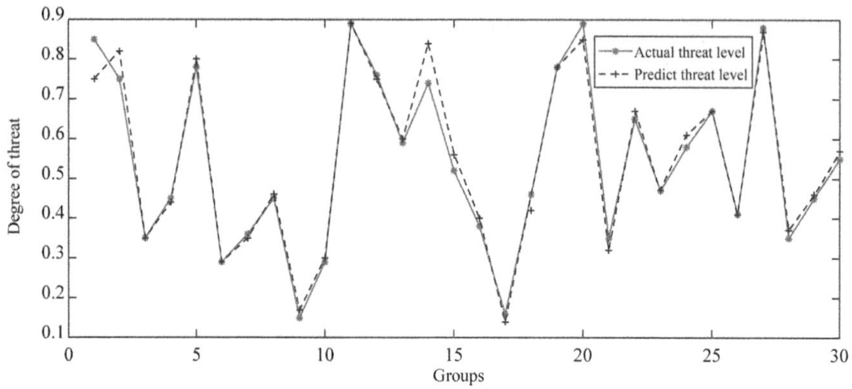

FIGURE 6.17 Comparison of predicted versus actual threat levels

For validation, the remaining 30 sets were used as an input to obtain the threat assessment under corresponding conditions and compared with the actual threats, as shown in Figure 6.17.

When the allowed error is less than 0.05, the alignment between the model's threat assessment results and the actual values is 86.6%. When the allowed error is less than 0.1, the alignment tends to achieve 96.6%. Therefore, the cognitive intelligent interactive human-machine cooperative threat assessment model has good practical application value.

6.4 CHAPTER SUMMARY

The main work completed in this chapter includes the following perspectives:

(1) Designed a task scenario for cognitive intelligent interactive cooperative decision-making between manned systems and UAVs.

(2) Incorporated the manned aircraft operator's cognitive workload into the cooperative situation awareness model design based on the Dynamic Intuitionistic Fuzzy Cognitive Map (DIFCM), reflecting the dynamic uncertainty of the external environment and the cognitive characteristics of the operator.

(3) Established a cooperative threat assessment model by combining the Intuitionistic Fuzzy Cognitive Map (IFCM) with the genetic algorithm, embodying the intelligent features driven by both expert knowledge and historical data.

CHAPTER 7

Cognitive Intelligent Interactive Human-Machine Cooperative Task Allocation and Path Planning Model

THIS CHAPTER DEVELOPS COOPERATIVE task allocation and path planning models for manned-unmanned teams. The models are built on a cognitive-intelligent interactive decision-making framework that accounts for the pilot's cognitive load and the UAV's intelligent affective mode.

7.1 COGNITIVE INTELLIGENT INTERACTIVE HUMAN-MACHINE COOPERATIVE TASK ALLOCATION MODEL

7.1.1 Analysis of Cooperative Task Allocation Problem

The purpose of manned-unmanned cooperative task allocation is to dynamically schedule UAV capability resources according to the battlefield situation. This approach improves UAV survivability, increases mission success rates, and enhances the overall effectiveness of the formation [96–99]. The processing flow of UAV task allocation is shown in Figure 7.1.

DOI: 10.1201/9781003791713-7

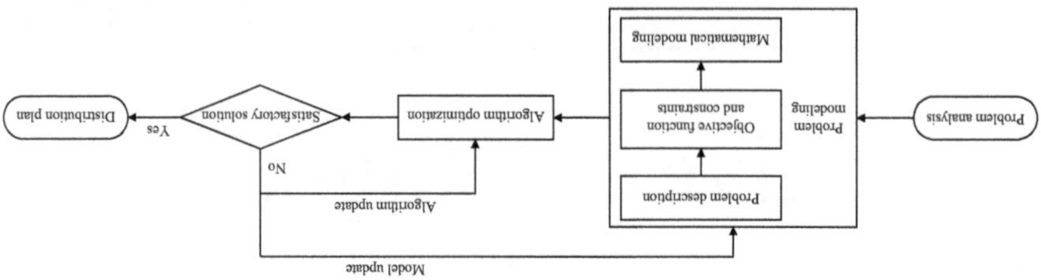

FIGURE 7.1 Task allocation processing flow of UAV

7.1.1.1 Problem Description

Problem description in task allocation involves defining the UAV task allocation process using mathematical language to clarify the specific variables, assumptions, and objective functions. Additionally, the problem complexity can be simplified. To describe the multi-UAV, multi-target task allocation problem, suppose there are n friendly UAVs $V_i\left(V_i \in V, i=1,2,\cdots,n\right)$ and t enemy targets $T_i(T_i \in T, i=1,2,\cdots,t)$. Let set $S_i=\left\{T_i^1,T_i^2,\cdots,T_i^n\right\}$ be the target set assigned to V_i.

7.1.1.2 Objective Function and Constraints

The objective function is generally defined based on specific operational requirements, often appearing as tactical indicators. Examples include maximum target profit, minimum friendly UAV damage, the shortest flight path, and minimum weapon cost consumption.

Constraints are derived from the problem description, defining the value ranges for task allocation parameters according to the UAV's flight and operational performance limits. This ensures that the objective function is optimized within feasible bounds. In practice, only the main factors are considered as constraints to reduce solving difficulties caused by excessive constraints.

7.1.1.3 Mathematical Model

Common mathematical models in task allocation problems include assignment problems, traveling salesman problems, and transportation problems. For UAV task allocation, there is no unified mathematical model, and it needs to be established according to the specific characteristics and requirements of the task allocation problem.

7.1.2 Structure of Cognitive Intelligent Interactive Cooperative Task Allocation Model

Existing research on human-machine cooperative task allocation models rarely considers the cognitive load state of the manned aircraft pilot within the formation or the impact of the UAV's state on cooperative task allocation. In the cognitive intelligent interactive cooperative task allocation problem, based on cooperative situation assessment and threat assessment, a cooperative task allocation structure adaptive to profit and threat is established by introducing the manned aircraft pilot's cognitive load state and the UAV's intelligent operational emotion mode, as demonstrated in Figure 7.2.

7.1.3 Cognitive Intelligent Interactive Cooperative Task Allocation Model

The contract net is used here as the basic modeling method for the cooperative task allocation model. The contract net simulates the "bidding–tendering–winning" mode in real-world market mechanisms, where multiple agents communicate and negotiate to complete task allocation. Agents represent individuals in the contract net, are capable of communicating and processing information, and possess a certain degree of autonomy. These agents search for global or near-global optima while pursuing individual optima. The contract net consists of these communicating agents,

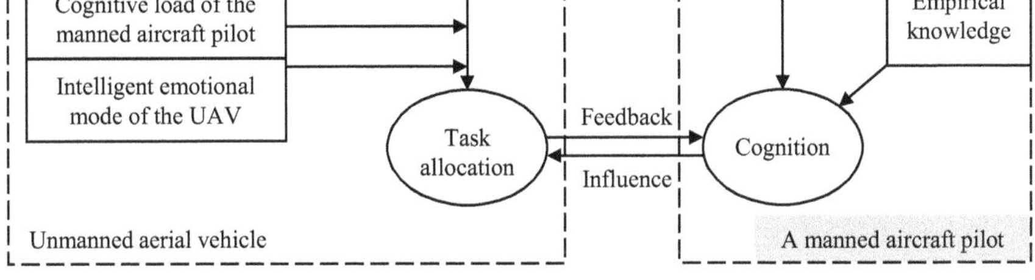

FIGURE 7.2 Task allocation processing flow of UAV

which can be divided into the following three types according to the market auction mechanism:

(1) Bidding Agent. The bidding agent owns the tasks, publishes tasks, analyzes and compares tenders from UAVs, and decides which UAV wins the task.

(2) Tendering Agent. A tendering agent has sufficient capability to complete the task published by the bidding agent.

(3) Winning Agent. The winning agent signs a contract with the bidding agent, obtaining ownership of the task published by the bidding agent.

The basic flow of the contract net task allocation algorithm is as follows: The bidding agent publishes task information. When other agents receive the task published by the bidding agent, they send bid information to the tendering agent within a limited time based on their capabilities. Within the specified bid deadline, the bidding agent analyzes and processes all bids from the tendering agents, selects the most suitable tenders, and sends the bid results to all tendering agents. After receiving the bid results returned by the bidding agent, the winning agent signs a contract with the bidding agent, adds the task to its own task sequence, prepares to execute the task, and returns the task completion status to the bidding agent after a period of time, as shown in Figure 7.3.

Assumption: one manned aircraft and N UAVs $V_i, i = 1,2,\cdots,N$ form a hybrid formation. There are M targets $T_j, j = 1,2,\cdots,M$ to be allocated. The target allocation matrix $X_{N\times M}$ is as follows:

$$x_{ij} = \begin{cases} 1, & T_j \text{ is assigned to } V_i \\ 0, & T_j \text{ isn't assigned to } V_i \end{cases} \tag{7–1}$$

The cooperative task allocation effectiveness function is as follows:

$$U(X) = E(X) - C(X) + D(X) \tag{7–2}$$

$U(X)$ is the effectiveness function for the manned aircraft and UAV formation task allocation, $E(X)$ is the revenue function for the manned aircraft and UAV formation task allocation. $C(X)$ is the cost function.

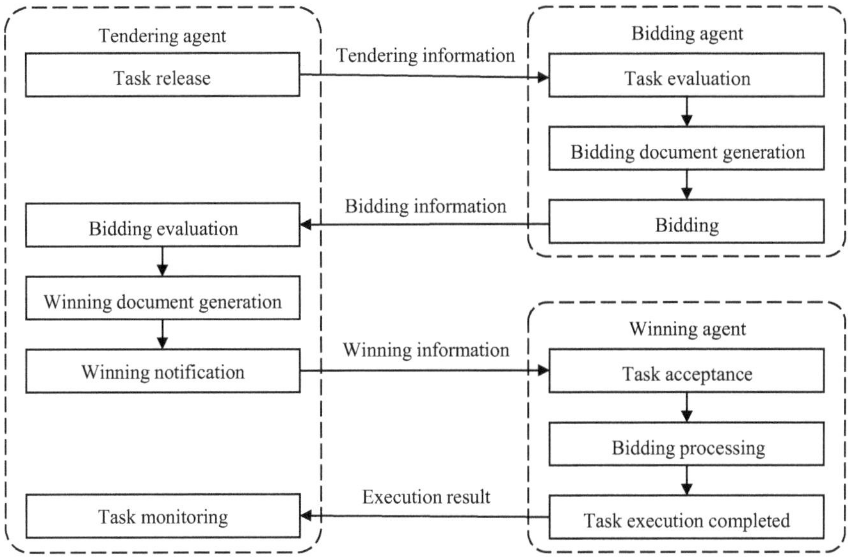

FIGURE 7.3 Flowchart of the contract net algorithm

$D(X)$ is the revenue adjustment function, dynamically adjusted according to the UAV's intelligent emotion mode. For instance, in the fear mode, increase the revenue adjustment function value for UAVs with smaller task revenues, making their task effectiveness function value higher and easier to assign the task. In the aggressive mode, increase the revenue adjustment function value for UAVs with higher task revenues, making their task effectiveness function value higher and making the task easier to assign.

Assumption: The formation's maximum revenue is E_{max}, the minimum revenue is E_{min}, and the difference between the maximum and minimum revenues is E_d. Let D_{ij} be the element in the i-th row and j-th column of the revenue matrix $D(X)$. The value of the adjustment function D_{ij} is determined as follows:

(1) Fear Mode:

$$D_{ij} = \begin{cases} \alpha \cdot E_{ij}, & E_{ij} \leq E_{min} + 0.3 \cdot E_d \\ 0, & \text{others} \end{cases} \tag{7-3}$$

(2) Relaxed Mode:

$$D_{ij} = \begin{cases} \alpha \cdot E_{ij}, \ E_{\min} + 0.3 \cdot E_d < E_{ij} \le E_{\min} + 0.7 \cdot E_d \\ 0, \quad \text{others} \end{cases} \tag{7–4}$$

(3) Aggressive Mode:

$$D_{ij} = \begin{cases} \alpha \cdot E_{ij}, \ E_{\min} + 0.7 \cdot E_d < E_{ij} \le E_{\max} \\ 0, \quad \text{others} \end{cases} \tag{7–5}$$

Here, α represents the preference coefficient, with a value range of [0.1, 0.5].

7.1.3.1 UAV Task Revenue Model

In the cognitive intelligent interaction mode, UAV task execution needs to consider the current cognitive load state of the manned aircraft operator. The task allocation revenue model can be expressed as follows:

$$E_u = Q_m \cdot f(\lambda) \cdot S_u \tag{7–6}$$

where E_u is the revenue of the UAV, Q_m is the reliability factor of the manned aircraft operator. $f(\lambda)$ is the influence factor of the manned aircraft pilot's cognitive load. S_u refers to the attack efficiency of the formation UAV.

The reliability factor Q_m of the manned aircraft operator is closely related to the task execution time, generally decreasing exponentially over time:

$$Q_m = e^{-\varepsilon(t-t_0)} \tag{7–7}$$

where ε is the failure rate, representing the average rate of errors made by the operator over time. t_0 is the task start time, and t is the current time.

The cognitive load influence factor $f(\lambda)$ is established by analyzing the manned aircraft pilot's cognitive load using a sigmoid function as follows:

$$f(\lambda) = \frac{1}{1+e^{-h\lambda}} \tag{7–8}$$

λ is the evaluated value from the manned aircraft operator cognitive load model, h is the slope of the sigmoid, controlling the degree of influence of the pilot's cognitive load on UAV effectiveness.

The attack efficiency S_u of the UAV represents the ideal revenue of the UAV striking the enemy target unit, which is expressed as follows:

$$S_u = V\_TARGET_j \cdot P_k \tag{7–9}$$

V_TARGET_j is the value of the enemy target j, and p_k is the probability of the UAV destroying the target. Hence, the UAV task revenue model, considering the manned aircraft operator's cognitive load state, is as follows:

$$E_u = \frac{V\_TARGET_j \cdot P_k \cdot e^{-\varepsilon(t-t_0)}}{1+e^{-h\lambda}} \tag{7–10}$$

7.1.3.2 UAV Task Cost Model

The cost of a UAV attacking an enemy target mainly considers: (1) the cost of the UAV being destroyed and (2) the threat level of the enemy target to our side. Let the target threat matrix be $THRAEAT\left(threat_{ij}, i \in N, j \in M\right)$, and the UAV task cost model is as follows:

$$C_{\text{u}} = k_1 \cdot V\_\text{UAV}_i \left(1 - \prod_{j=1}^{m}(1-P_j)\right) + k_2 \cdot \text{threat}_{ij} \tag{7–11}$$

C_u is the cost of the UAV executing the task. k_1 and k_2 are weight coefficients. V_UAV_i represents the own value of the UAV i. P_j represents the probability of the target j destroying the UAV. m is the number of targets. $threat_{ij}$ is the threat degree of the target j to the UAV i. In summary, the effectiveness function for cooperative task allocation is as formulated in 7.12.

$$U_u = E_u - C_u + D_{ij} \tag{7–12}$$

7.1.4 Case Simulation Analysis

Assume that one manned aircraft and four UAVs perform a cooperative attack task allocation for six ground targets. The battlefield situation is shown in Figure 7.4, and the information on the six targets is shown in Table 7.1.

Calculate the target threats using the cooperative threat model from Chapter 6 and set the target values, as shown in Table 7.2.

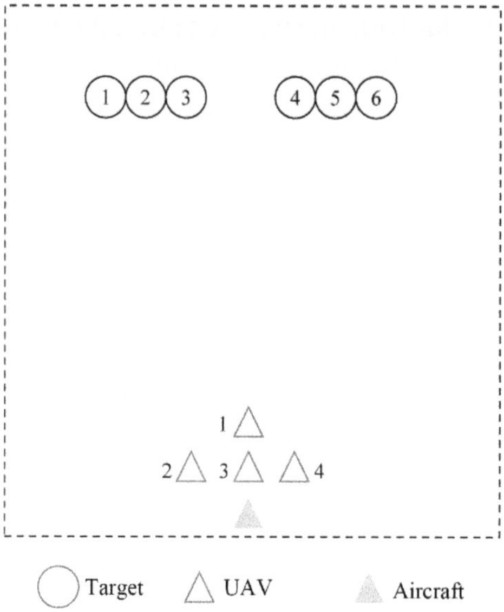

FIGURE 7.4 Schematic diagram of the initial battlefield situation

TABLE 7.1 Enemy Ground Target Information

Target Number	Type	Position Threat	Maneuverability	Electronic Countermeasure Capability	Weapon's Destructive Capability	Radar Detection Capability	Cognitive Load
1	0.1	0.5	0.4	0.6	0.47	0.6	0.5
2	0.1	0.5	0.5	0.52	0.5	0.7	0.5
3	0.3	0.7	0.6	0.4	0.5	0.72	0.5
4	0.3	0.7	0.6	0.4	0.4	0.69	0.5
5	0.5	0.8	0.85	0.75	0.76	0.6	0.5
6	0.5	0.8	0.7	0.85	0.75	0.76	0.5

TABLE 7.2 Value and Threat Level of Enemy Targets

Target Number	Target Value	Degree of Threat
1	0.8	0.3486
2	0.3	0.3756
3	0.5	0.5125
4	0.6	0.5048
5	0.4	0.7839
6	0.8	0.8125

Among them, Target 1 has low threat and high value, Target 2 has low threat and low value, Targets 3 and 4 have medium threat and medium value, Target 5 has high threat and low value, and Target 6 has high threat and high value.

Simulation analysis based on the cognitive-intelligent interactive task allocation model identifies UAV intelligent affective modes that consider the operator cognitive load. The corresponding task allocation results are shown in Table 7.3.

Simulation analysis shows that Target 1 is prioritized due to its low threat and high value, and Target 5 is basically not allocated due to its high threat and low value. The threat level and total value of the target allocation results under various UAV intelligent emotion modes are summarized, and the results are shown in Figure 7.5.

TABLE 7.3 Task Allocation Results Under Different Intelligent Emotion Modes

Types of UAV Emotions	UAV 1	UAV 2	UAV 3	UAV 4
Fear	target 1	target 2	target4	target3
Relaxed	target 1	target 3	target4	target2
Aggressive	target 1	target 6	target4	target3

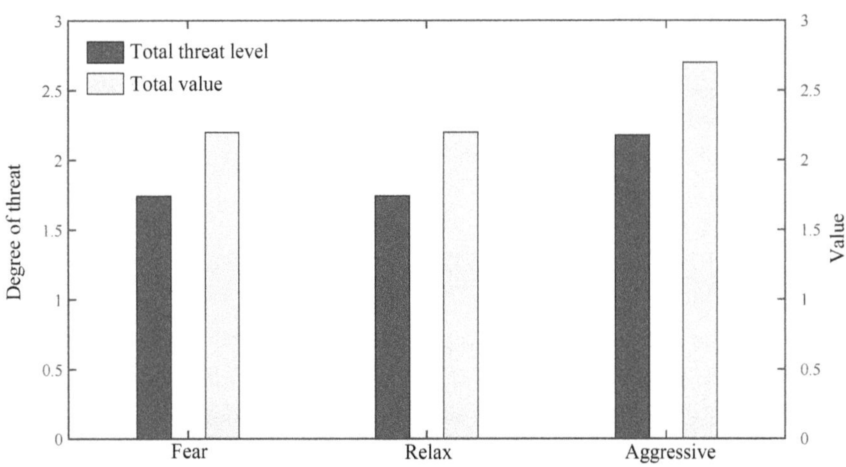

FIGURE 7.5 Threat and value of target allocation under different modes

Analysis shows that when UAVs are in the fear or relaxed intelligent emotion modes, they tend to prefer targets with low threat and not high value. When UAVs are in aggressive intelligent emotion mode, they choose targets with high threat and high value, and this verifies the influence of the manned aircraft operator's cognitive load and UAV intelligent emotion on task allocation.

7.2 COGNITIVE INTELLIGENT INTERACTIVE HUMAN-MACHINE COOPERATIVE PATH PLANNING MODEL

7.2.1 Analysis of Cooperative Path Planning Problem

For the cognitive intelligent interactive cooperative path planning problem between manned aircraft and UAVs, the purpose is to find a reasonable flight path from the starting point to the endpoint for the UAV to avoid various threats and successfully reach the designated mission area. Similarly, the cognitive load of the manned aircraft operator and the intelligent emotion mode of the UAV need to be considered, which are crucial for planning UAV flight paths that adapt to the characteristics of the cognitive intelligent interaction mode.

Currently, a great deal of research has been conducted on path planning algorithms domestically and internationally, which can be broadly divided into two categories: graphics-based path planning algorithms and intelligent optimization-based path planning algorithms. Graphics-based algorithms can be divided into topological algorithms, grid algorithms, and artificial potential field algorithms. Intelligent optimization algorithms include swarm intelligence optimization algorithms and bionic intelligence optimization algorithms, and mainly include the A* algorithm, genetic algorithm, ant colony optimization algorithm, particle swarm optimization algorithm, to name a few [100, 101].

7.2.2 Structure of Cognitive Intelligent Interactive Cooperative Path Planning Model

To ensure the safety of UAV flight paths and the reliability of mission execution, when the cognitive load of the manned aircraft operator is high, the UAV's flight path should be relatively far from threat points to ensure flight safety. When the cognitive load of the manned aircraft operator is low, the flight path can be relatively closer to threat points to ensure rapid mission execution. Therefore, the cognitive intelligent interactive cooperative path planning model needs to combine the cognitive load state of the manned aircraft operator, battlefield threats, and other information to dynamically switch the UAV's intelligent emotion mode, thereby influencing the path-planning process, as shown in Figure 7.6.

During cooperative path planning, if a sudden threat appears on the UAV's flight path, the response depends on the operator's cognitive load. If the cognitive load is low, the operator can manually set waypoints to adjust the path. If the cognitive load is high, the UAV autonomously plans its route based on its current intelligent affective mode and selects appropriate decision preferences.

7.2.3 Cognitive Intelligent Interactive Cooperative Path Planning Model

The A* approach, as a heuristic search algorithm, has the advantages of simplicity and ease of implementation. Its main idea is to continuously calculate the cost value of the next adjacent node from the initial point based on the designed heuristic function, compare the cost values of adjacent nodes, select the node with the smallest cost as the next node to be expanded, and continue the search and comparison until the target point is reached. The entire search process obtains a set of nodes,

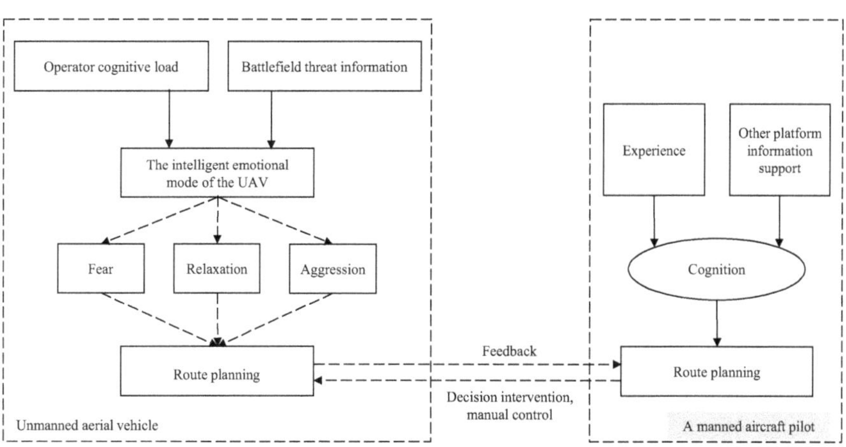

FIGURE 7.6 Structure of cognitive intelligent interactive human-machine cooperative path planning

and connecting these nodes in order yields the desired optimized route. The path-planning model based on the A* algorithm is introduced in the following.

(1) Design and establish two data tables: an open table and a close table. The open table stores points whose costs have been calculated but have not been expanded, and the close table stores nodes that have been expanded. The path node stores multiple pieces of information as follows:

$$\text{Node} = \{(x_p, y_p), (x_i, y_i), g(i), h(i), f(i)\} \tag{7–13}$$

(x_p, y_p) is the coordinate of the parent node of the i -th node in the UAV path. (x_i, y_i) is the coordinate of the i -th waypoint. $g(i)$ is the path cost from the waypoint i to the starting point. $h(i)$ is the estimated cost from the UAV at the waypoint i to the target point. $f(i)$ is the minimum path cost to waypoint i. The corresponding calculation formulas are as follows:

$$f(i) = g(i) + h(i) \tag{7–14}$$

$$g(i) = \sum_{j=1}^{i} (k_1 l_j + k_2 w_j) \tag{7–15}$$

k_1 and k_2 are weight coefficients; k_1 is generally taken as 1; k_2 affects the threat cost. The larger the value is, the farther the path is from the threat points. l_j is the distance between the j -th node and its parent node. w_j is the threat cost suffered by the j -th node, which is defined as follows:

$$w_j = \sum_{k=1}^{M} \frac{\delta}{d_{jk}^2} \tag{7–16}$$

M represents the number of enemy obstacles or threats. δ is the threat coefficient, indicating the degree of threat. d_{jk} denotes the distance from the j -th route point to the center of the enemy obstacle or threat point k.

$$h(i) = \sqrt{(x_n - x_i)^2 + (y_n - y_i)^2} \tag{7–17}$$

(x_n, y_n) is the position of the target point, and (x_i, y_i) represents the position of the i -th waypoint.

(2) Add the starting node to the "close" table and set it as the current node.

(3) Search for the neighboring nodes of the current node, calculate the route value $f(i)$ of the neighboring nodes, and add them to the open list. During the iteration process, find the node with the smallest value from the open list, add it to the close list, and delete this node from the open list.

(4) If the selected node is the target node, the route-planning process will be terminated. Recursively search for the parent nodes of this node until reaching the starting node; obtaining the route nodes is expressed as follows:

$$path = \{(x_0, y_0), (x_1, y_1), \cdots, (x_n, y_n)\} \tag{7–18}$$

(5) If the selected node is not the target node, then it is regarded as the current node, and the process proceeds to Step (3) until the target point is found. The flowchart of the A* algorithm is shown in Figure 7.7.

Based on the cognitive intelligent interactive human-machine collaborative route planning model structure, the A* algorithm was improved. When the unmanned aircraft conducts route planning, it adjusts the obstacle avoidance radius according to the current intelligent emotional mode. The avoidance radius is defined as the minimum distance to the center of the threat point, and it is shown in Figure 7.8.

The formula for calculating the avoidance radius D is as follows:

$$D = R \cdot e^{(1-1/R)\cdot(1-C_w)} \tag{7–19}$$

R represents the threat radius, and C_w represents the intelligent emotional mode of the UAV, which includes three modes: fear, relaxation, and aggression. These modes are determined by the parameters α and β as follows:

$$\begin{cases} Fear, & C_w < \alpha \\ Relaxed, & \alpha \le C_w \le \beta \\ Aggressive, & C_w > \beta \end{cases} \tag{7–20}$$

where the range of α and β is $0 < \alpha < \beta < 1$.

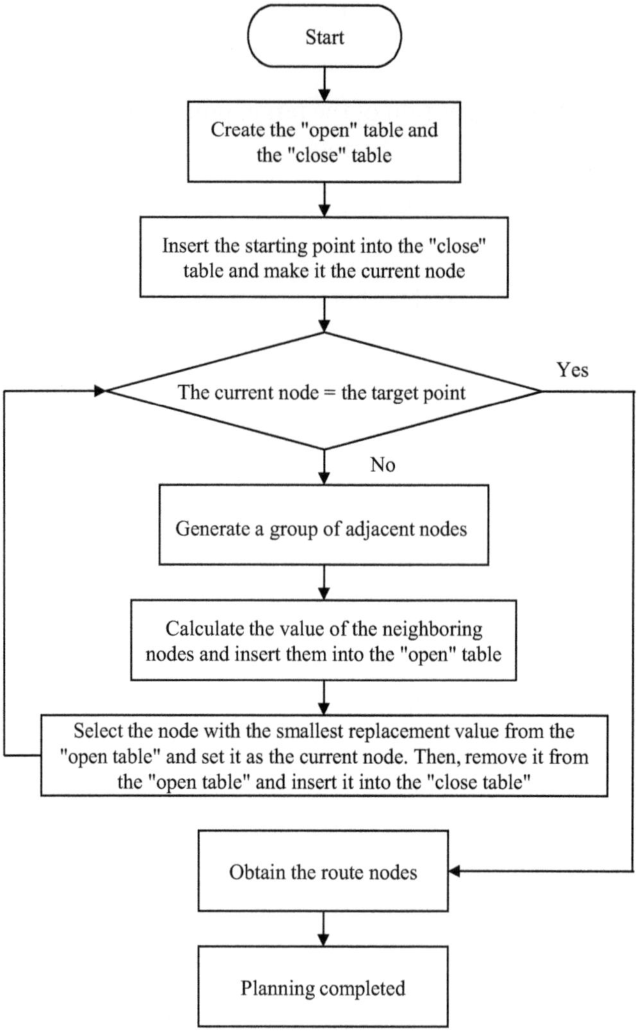

FIGURE 7.7 A* algorithm route planning flowchart

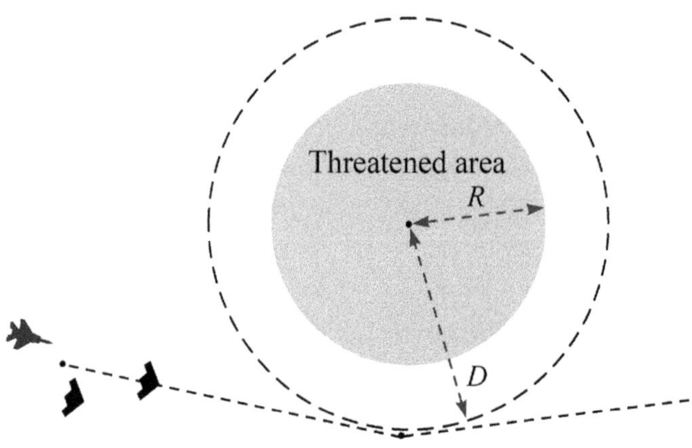

FIGURE 7.8 Threat avoidance radius

Under different emotional states, the avoidance radius D for the drone to switch to other modes is calculated according to the following formula.

$$\begin{cases} Fear, & Current\ mood \\ D = R\cdot e^{(1-1/R)\cdot[1-\frac{C_w}{\alpha}(\beta-\alpha)-\alpha]}, & Relaxed \\ D = R\cdot e^{(1-1/R)\cdot[1-\frac{C_w}{\alpha}(1-\beta)-\beta]}, & Aggressive \end{cases} \tag{7-21}$$

$$\begin{cases} Relaxed, & Current\ mood \\ D = R\cdot e^{(1-1/R)\cdot(1-\frac{C_w-\alpha}{\beta-\alpha}\cdot\alpha)}, & Fear \\ D = R\cdot e^{(1-1/R)\cdot[1-\frac{C_w-\alpha}{\beta-\alpha}(1-\beta)-\beta]}, & Aggressive \end{cases} \tag{7-22}$$

$$\begin{cases} Aggressive, & Current\ mood \\ D = R\cdot e^{(1-1/R)\cdot(1-\frac{C_w-\beta}{1-\beta}\alpha)}, & Fear \\ D = R\cdot e^{(1-1/R)\cdot[1-\frac{C_w-\beta}{1-\beta}(\beta-\alpha)-\alpha]}, & Relaxed \end{cases} \tag{7-23}$$

7.2.4 Case Simulation Validation

Figure 7.9 shows the human-machine interaction interface for cognitive intelligent interactive cooperative path planning. The left side is the battlefield situation map, mainly displaying the relative positions of manned aircraft, UAVs, targets, and threat sources, as well as showing the planned flight routes of the UAVs. The black dashed line represents the planned route when the UAV intelligent emotion mode is in the aggressive mode. The thick white dashed line represents the planned route in the relaxed mode. The thin white dashed line represents the planned route in the fear mode. The manned aircraft can click "Threat Information" to set or modify threat point information. The UAV can autonomously plan a suitable flight path based on its current intelligent emotion mode. If the manned aircraft operator is dissatisfied with the route planned by the UAV, they can directly specify the UAV to fly according to a specific route by selecting "Select Route". When the UAV faces a sudden threat and the re-planned route deviates too much from the preplanned route, the manned aircraft operator can set guide points to assist the UAV in optimizing the route as much as possible according to the preplanned route. The right part of the

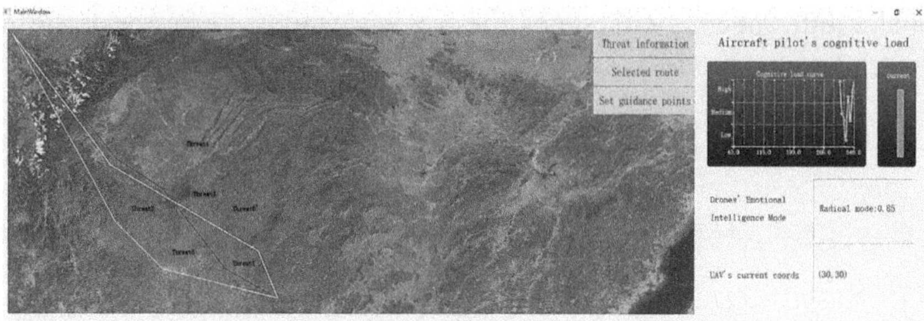

FIGURE 7.9 Cognitive intelligent interactive collaborative route planning human-computer interaction interface

human-machine interaction interface displays the current cognitive load of the manned aircraft operator and the intelligent emotion mode of the UAV. The cognitive load of the manned aircraft operator is displayed as a line chart.

The following section demonstrates the critical information in the cooperative path planning human-machine interaction interface:

7.2.4.1 Threat Information

Considering sensor or weather-related reasons, the acquired threat information may not be accurate enough. To ensure the flight safety of the UAV and the smooth execution of combat missions, the manned aircraft operator can add or modify the UAV's threat information based on the actual situation, the information mastered by the manned aircraft, and their own cognitive judgment, as shown in Figure 7.10.

The manned aircraft operator can click the "Add" button to add threat information for a specific UAV. The known threat information currently obtained by the UAV appears in the text box in the lower-left corner of Figure 7.10. If a threat is destroyed but not updated due to communication interruption or delay, it can be deleted using the "Delete" button. Clicking the "Update Route" button actively sends instructions requesting the UAV to replan the route. The manned aircraft operator can select one of the three autonomous routes as the predefined flight path, which automatically deletes the other two options. Alternatively, the operator can choose "Select Route" to manually plan and assign a custom flight path for the UAV to follow.

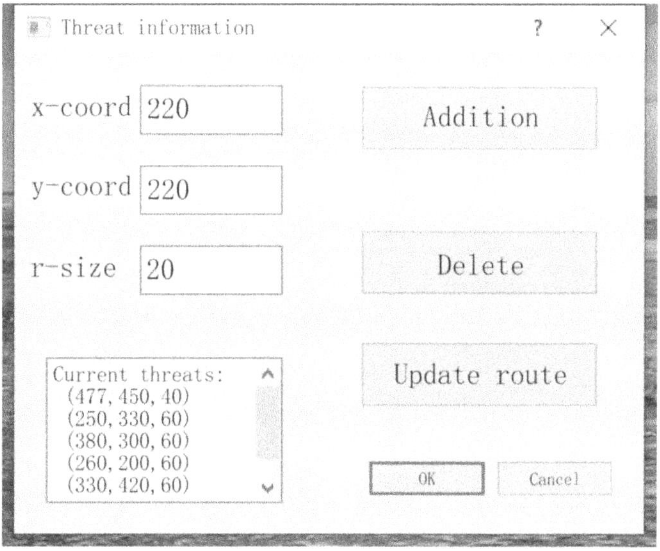

FIGURE 7.10 Threat information interface

7.2.4.2 Cognitive Load and Intelligent Emotion Mode

The cognitive load data of the manned aircraft operator is displayed as a line chart. The cognitive load level is displayed in the histogram on the right side of the line chart and is updated periodically. If the operator's cognitive load is too high, the histogram turns red to alert the operator. The curve on the left side of the line chart represents the operator's historical cognitive load. The intelligent emotion mode of the UAV is displayed numerically.

7.2.4.2.1 Simulation Validation of Cooperative Path Planning with Known Threats To facilitate analysis of the path planning effects under different UAV intelligent emotion modes, a set of battlefield threat information was established, reducing the influence of other factors on path planning. The simulated situation map is a rectangular area of 800×600. The initial threat information is shown in Table 7.4.

The results of the drone route planning under the three intelligent emotional modes are shown in Figure 7.11. Here, the black dotted line represents the route in the aggressive mode, the white thick dotted line represents the route in the relaxed mode, and the white thin dotted line represents

TABLE 7.4 Enemy Threat Information

Threat Point	1	2	3	4	5	6
Center coordinates	(477, 450)	(250, 330)	(380, 300)	(360, 200)	(330, 420)	(480, 340)
Threat radius	40	60	60	60	40	40

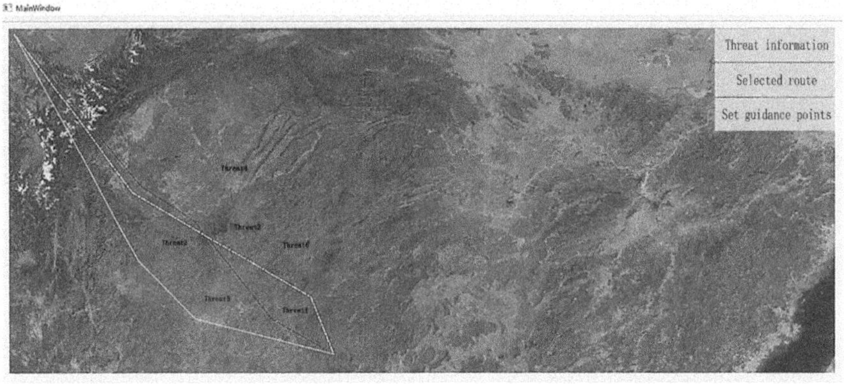

FIGURE 7.11 Route planning result

the route in the fearful mode. It can be seen that when the drone is in the aggressive mode, the planned route is the shortest to the target, and the overall flight route is closer to the threats, which is in line with the setting of the aggressive mode. In the fearful mode, the drone plans the route by avoiding all threats directly from below, ensuring the safety of the drone to the greatest extent, which is in line with the setting of the drone's fearful mode. In the relaxed mode, the distance between the drone's planned route and the threats is between that of the aggressive mode and the fearful mode. The detailed route planning situation is shown in Table 7.5.

The waypoints are the key nodes selected after the smooth processing of the flight path. As shown in Table 7.5, under the aggressive mode, the total flight distance of the unmanned aircraft is the shortest, and the threat cost is the highest. Under the fearful mode, the total flight distance is the longest, and the threat cost is the lowest, and the relaxed mode is between the two.

7.2.4.2.2 Simulation Validation of Cooperative Path Planning with Sudden Threats The unmanned aircraft flies along the preplanned route. If a sudden threat is detected, it can autonomously plan an avoidance route based

TABLE 7.5 The Route Planning Results Under the Three Modes

Mode	Navigation Point Information	Total Voyage	Cost of Threat
Aggressive	(10, 10), (227, 262), (321, 321), (394, 411), (453, 484), (560, 542)	774.99	0.349
Relaxed	(10, 10), (196, 225), (213, 273), (426, 394), (520, 447), (560, 542)	791.17	0.227
Fear	(10, 10), (156, 271), (223, 387), (321, 483), (560, 542)	816.38	0.165

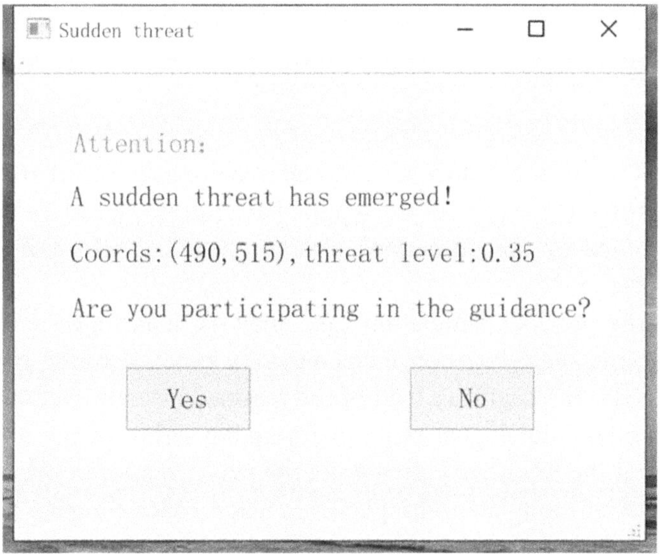

FIGURE 7.12 Sudden threat to human-computer interaction window

on the current intelligent emotional mode. If the pilot of the manned aircraft is not satisfied with the route planned by the unmanned aircraft, they can set guiding points through the human-machine interaction interface.

As shown in Figure 7.12, when a sudden threat occurs in the simulation, the human-machine interaction interface displays a threat information window. After remaining for one minute, it automatically exits. The pilot of the manned aircraft can choose "No" to ignore this information and let the unmanned aircraft plan the route autonomously or choose "Yes" to assist the unmanned aircraft in avoiding the sudden threat by setting guiding points.

As shown by the white dotted line in Figure 7.13, the unmanned aircraft's planned route bypasses the threat point from below because there

FIGURE 7.13 Emergency threat route planning

is a threat above the sudden threat, and avoiding the threat from above is more costly. If the pilot of the manned aircraft is not satisfied with the route planned by the unmanned aircraft, they can set guiding points (white dots).

From the previous simulation case analysis, it is suggested that the cognitive intelligent interactive collaborative route planning model can fully consider the cognitive load of the manned aircraft operator, utilize the autonomous intelligent route planning algorithm of the unmanned aircraft and the cognitive judgment of the manned aircraft operator, and collaboratively complete route planning tasks under complex battlefield environments.

7.3 CHAPTER SUMMARY

The main tasks accomplished in this chapter are as follows:

(1) Based on the contract network task allocation model, the cognitive load of human operators and the intelligent emotional patterns of unmanned aircraft were introduced, enhancing the dynamic adaptability of the collaborative task model under the cognitive intelligent interaction mode.

(2) The intelligent emotional patterns of unmanned aircraft were introduced into the collaborative route planning model, which improved the ability of unmanned aircraft to actively adapt to the cognitive load of human operators and autonomously plan feasible routes under complex and dynamically changing task environments.

CHAPTER 8

Cognitive Intelligent Interactive Simulation System for Manned Aircraft and UAVs Collaborative Decision-Making

THIS CHAPTER PRESENTS THE architecture of the cognitive intelligent interactive simulation system for manned aircraft and UAV collaborative decision-making, interprets the software and hardware implementation of its constituent subsystems in detail, and validates the simulation system's reliability and usability.

8.1 ARCHITECTURE OF THE COGNITIVE INTELLIGENT INTERACTIVE HUMAN-MACHINE COLLABORATIVE DECISION-MAKING SIMULATION SYSTEM

Simulation of human-machine collaborative decision-making can be divided into two parts: digital simulation and virtual-visual simulation.

DOI: 10.1201/9781003791713-8

Digital simulation involves designing software workflows, writing code, completing program testing, and validating the methodologies based on various collaborative decision-making models. Virtual-visual simulation is an immersive, interactive method based on computer graphics that provides the manned aircraft operator with visualized information. It integrates visual, auditory, and haptic feedback to create a virtual experimental simulation environment. Traditional digital simulation is limited to providing data in singular formats or simple graphical representations of experimental results. It is primarily used for static simulation, offline statistical analysis, and verification. By integrating human-in-the-loop digital simulation with virtual-scene simulation within scenarios designed for manned-unmanned teaming and full-mission elements, this approach can effectively assess the operator's cognitive state, simulate UAV intelligent decision-making, and model human-machine interaction behavior during simulation experiments. This provides powerful support for the experimental simulation and verification analysis of the methodologies related to cognitive intelligent interactive collaborative decision-making between manned aircraft and UAVs.

The architecture of the cognitive intelligent interactive human-machine collaborative decision-making simulation system is illustrated in Figure 8.1.

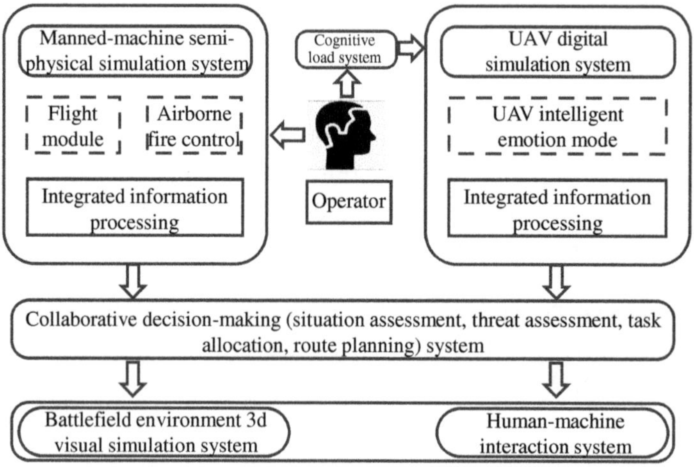

FIGURE 8.1 Architecture of the cognitive intelligent interactive human-machine collaborative decision-making simulation system

It primarily consists of the manned aircraft hardware-in-the-loop (HIL) simulation system, the UAV digital simulation system, the operator and cognitive workload system, the collaborative decision-making system, the 3D battlefield environment visual simulation system, and the human-computer interaction system. The experimental equipment is listed in Table 8.1.

The manned aircraft HIL simulation system comprises a full aircraft cockpit setup, including a throttle lever on the left, a control joystick on the right, and a large multifunction touchscreen display in front of the seat. A virtual reality–based 3D battlefield visual simulation is presented via six 50-inch monitors, providing subjects with a high level of immersion.

In addition, an eye tracker and a physiological monitoring device are included to support the cognitive workload assessment system, as shown in Figure 8.2. The selected performance specifications of the equipment are provided in Table 8.2.

TABLE 8.1 Experimental Equipment

Serial Number	Equipment Name	Quantity
1	50-inch monitor	6
2	Touch display	3
3	Lenovo workstations	4
4	Cockpit control system	1
5	Dikablis eye tracker	1

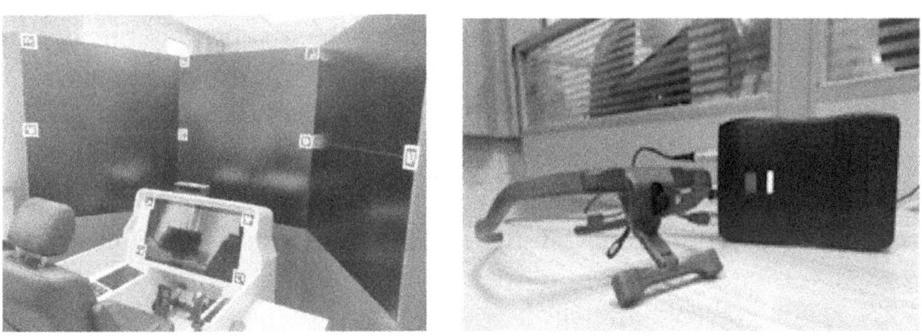

(a) Aircraft cockpit system (b) Eye tracker

FIGURE 8.2 Related simulation experimental equipment

TABLE 8.2 Some Equipment Performance Information

Equipment Type	Performance Information
Monitor	Resolution: 1280 × 1024, Average brightness: 120cd/m^2
Eye tracker	Sampling rate: 60Hz, Pupil tracking accuracy: 0.05°, Eye tracking accuracy: 0.1°~0.3°, Scene camera resolution: 1920 × 1080 Full HD, Scene camera range: 40°~90°

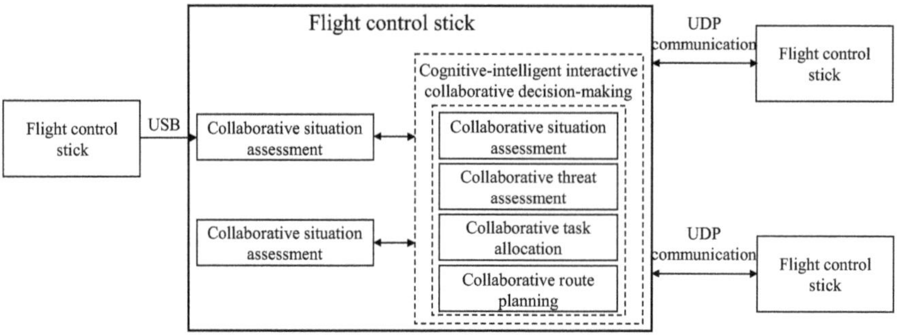

FIGURE 8.3 Information interaction architecture of the flight control simulation system

8.2 MANNED AIRCRAFT AND UAV COLLABORATIVE FLIGHT CONTROL SIMULATION SYSTEM

The flight control simulation system serves as the master control system for the collaborative flight of manned aircraft and UAVs. Within this system, the initial setup for collaborative operations is configured by specifying the initial positions of the manned aircraft and UAV, the types and quantities of carried weapons, enemy target information, threat data, and other parameters. Upon flight control system activation, the manned aircraft operator flies the aircraft via the control joystick and, in coordination with the UAVs, executes the combat mission. The flight control simulation system transmits the data to other subsystems for subsequent processing, utilization, and display. The information interaction architecture of the manned aircraft and UAV flight control simulation system is shown in Figure 8.3.

8.2.1 Flight Control Simulation System Software Interface

The flight-control simulation system's interface is developed using Visual C++ 6.0 with MFC. Before running a simulation, the manned aircraft's initial state must be configured on the initial-settings interface, typically set to "Take-off" status. If "Airborne" is selected, the initial altitude and Mach number of the manned aircraft are specified in the lower part of the interface. Based on simulation requirements, the user selects the subsystems to which data will be transmitted. In typical use, all subsystems are selected (e.g., sending simulation data to the visual scene system and the integrated situation awareness system), as shown in Figure 8.4. The UAV simulation software setup interface comprises modules for configuring the initial positions of both friendly and enemy forces, the supervisory control mode, threat information, and other parameters, as shown in Figure 8.5.

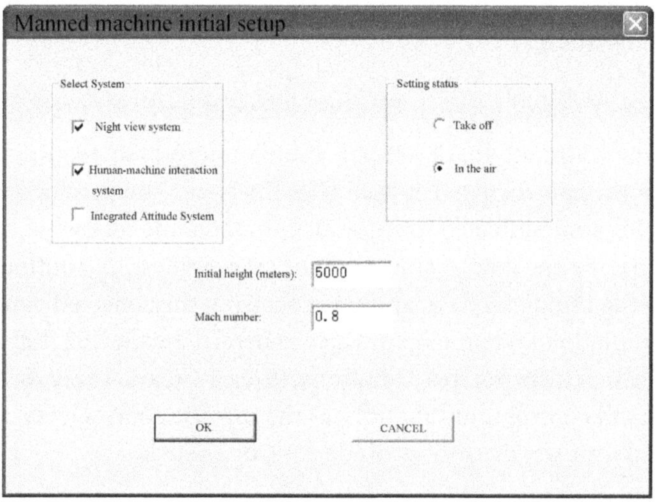

FIGURE 8.4 Manned aircraft initial setup interface

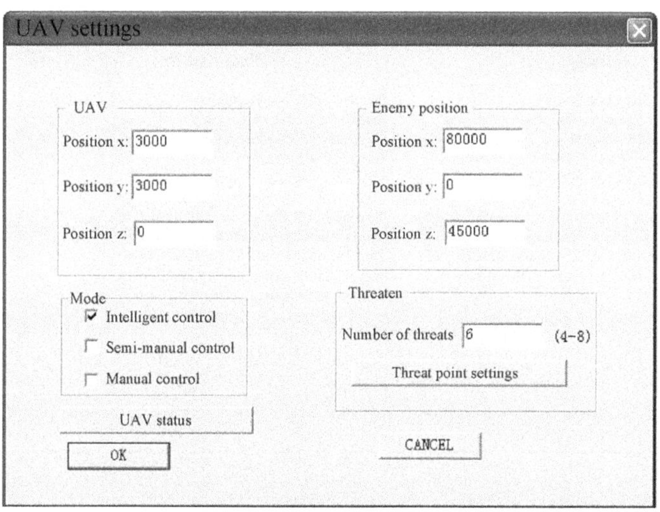

FIGURE 8.5 UAV initial setup interface

Other relatively fixed parameters of the simulation system are typically embedded within the simulation software code and databases. After completing the initial setup, the flight control simulation system software interface (Figure 8.6) displays friendly force operational information on the left side. This includes the number, speed, heading, and initial positions of UAVs, as well as whether airborne early warning (AEW) support is available. The right side of the interface displays enemy operational information, including the number and speed of hostile targets. Clicking the "Simulation Run" button initiates the operation of the flight control simulation system.

8.2.2 Flight Control Joystick Data Acquisition and Processing

The manned aircraft flight control simulation system requires reading data from the control joystick via a USB interface. The development work for the USB-compatible control joystick involves the following key steps: (1) determining the data protocol for the control joystick's button signals, (2) acquiring the signal data, and (3) processing the collected data.

The simulation system employs a commercial Saitek X52 flight control joystick, which incorporates 32 buttons, three axes, and three rotary dials. The left-hand control unit functions as the throttle control lever, while the right-hand unit serves as the attitude control joystick.

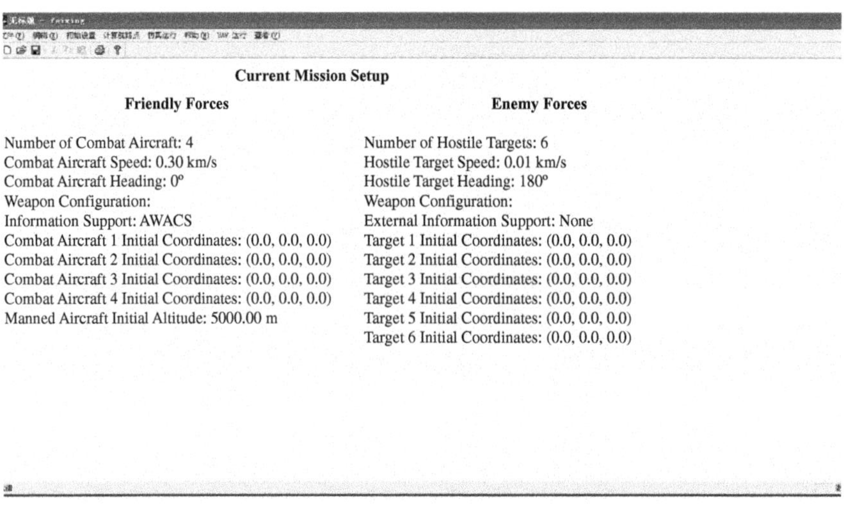

FIGURE 8.6 Flight control simulation system software interface

TABLE 8.3 USB Joystick Button-Signal Protocol

Key number	**1**	**2**	**3**	**...**	**31**	**32**
Key value	0X1	0X2	0X4	...	0X40000000	0X80000000

TABLE 8.4 USB Joystick Axis-Signal Protocol

Shaft Serial Number	***X*-axis**	***Y*-axis**	***Z*-axis**
Shaft value	0~65535	0~65535	0~65535

8.2.2.1 USB Control Joystick Protocol

A protocol is a set of rules governing information exchange between hardware devices and their corresponding programs in a system, analogous to an operational specification for the device. The USB control joystick button signal protocol is detailed in Table 8.3. When a button is triggered, the value of its corresponding 32-bit button code changes accordingly.

For the axes of the USB control joystick, the sampled data ranges from 0 to 65,535. This means when an axis is fully deflected in the positive direction, the sampled value is 65,535, and when it is fully deflected in the negative direction, the value is 0. The simulation system processes the corresponding data based on this setup. The protocol for the USB control joystick axes is detailed in Table 8.4.

8.2.2.2 Data Acquisition

USB joystick data acquisition can be accomplished by calling the built-in Windows WIN API. The Windows operating system provides a dedicated system service for monitoring the state of USB control joysticks, allowing direct reading of axis rotation data and button information through the corresponding functions. The relevant functions for the USB control joystick are packaged within Winmm.dll, with the corresponding header file being Mmsystem.h. The associated information for the USB control joystick is encapsulated within the JOYINFOEX structure.

```
typedef struct joyinfoex_tag{
DWORD dwSize,
DWORD dwFlags,
DWORD dwXpos,
DWORD dwYpos,
DWORD dwZpos,
DWORD dwRpos,
```

```
DWORD dwUpos,
DWORD dwVpos,
DWORD dwButtons,
DWORD dwButtonNumber,
DWORD dwPOV,
DWORD deReserved1,
DWORD deReserved2}
JOYINFOEX
```

Key variable information in the JOYINFOEX structure are as follows:

(1) dwSize specifies the size of the structure in bytes.

(2) dwXpos indicates the position data of the *x*-axis, dwYpos indicates the position data of the *y*-axis, and dwZpos indicates the position data of the *z*-axis.

(3) dwButtons represents the joystick button state data.

By defining a variable of the JOYINFOEX structure type and utilizing the corresponding acquired data, the application completes the acquisition and processing of joystick information.

8.2.3 UDP Communication of the Simulation System

The flight control simulation system communicates with other subsystems using the UDP protocol. UDP does not require a connection to be established before communication, offering advantages in speed and low resource consumption. Nevertheless, it carries the risk of packet loss. In addition, for the visual effects required by the visual scene system, the primary concerns are real-time performance and data refresh rate, while transmission accuracy is less critical, and occasional packet loss can be disregarded.

Under the UDP protocol, the communication program is developed using socket programming. The socket sends data to the driver and onto the network. Other simulation subsystems receive information from the corresponding IP addresses bound to these sockets. The network communication flowchart is shown in Figure 8.7.

8.3 VISUAL-SCENE SIMULATION SYSTEM

The three-dimensional battlefield visual-scene simulation system is developed using Vega Prime (VP). VP is an efficient virtual reality development

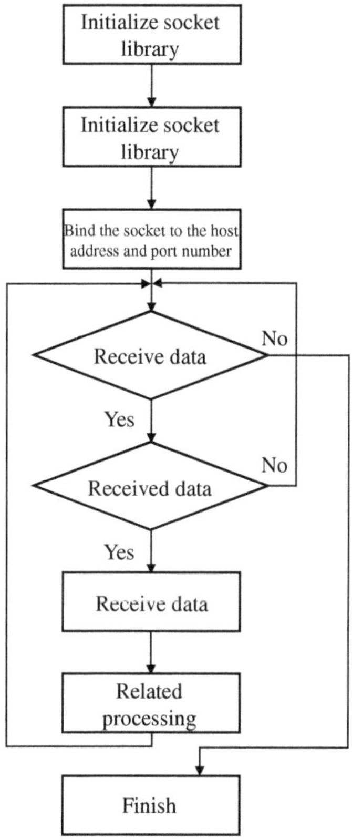

FIGURE 8.7 UDP network communication flowchart

software tool. Utilizing the Lynx Prime (LP) interface for application development, programmers can configure required functionalities through LP or invoke the corresponding Application Programming Interfaces (APIs) to achieve specific objectives [102]. The architecture of the Vega Prime system is shown in Figure 8.8. Owing to Vega Prime's strong applicability and extensibility, it significantly reduces the time and effort developers spend on 3D visual scene development, which offers the following main advantages:

(1) It features a graphical application development tool (i.e., LP), which requires minimal programming, and virtual scenes can be easily constructed using LP alone.

(2) It incorporates numerous modules, such as ocean, motion, special effects, and collision modules. These allow for the configuration

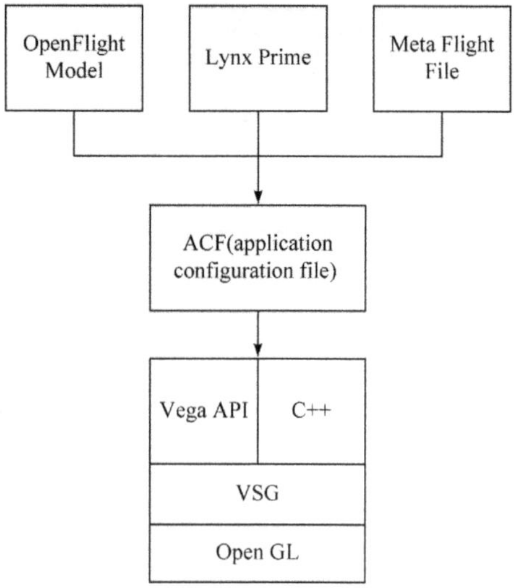

FIGURE 8.8 Vega Prime system structure

and development of customized visual systems tailored to specific requirements.

(3) It is cross-platform, supporting both Windows and Linux systems. This enables a "develop once, deploy multiple times" mechanism. It has strong compatibility and allows for program design using template functions.

(4) LP includes lighting and atmospheric models, along with specialized texture mapping techniques.

The development workflow of the Vega Prime–based 3D battlefield environment visual simulation system is illustrated in Figure 8.9, primarily including the design of object models within the visual scene and the definition of the Application Configuration File (ACF).

Figure 8.10 illustrates the execution workflow of the visual simulation program based on Vega Prime. The specific steps are as follows:

(1) Program Initialization: This primarily involves initializing static variables, rendering libraries, and module interfaces within Vega Prime.

(2) Variable Definition: This step mainly consists of parsing the ACF, as well as defining the entity classes within it.

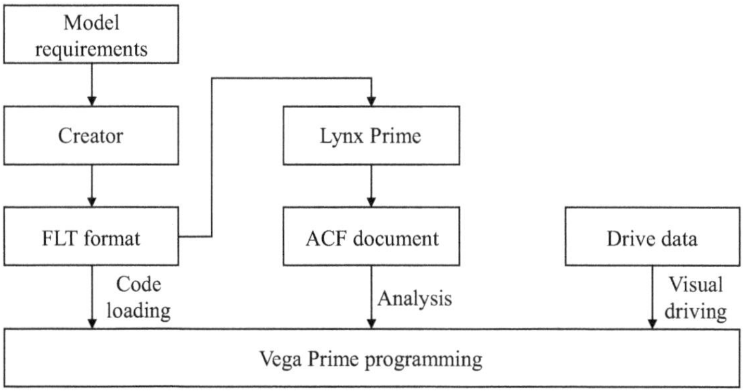

FIGURE 8.9 Development workflow of the Vega Prime–based 3D battlefield environment visual simulation system

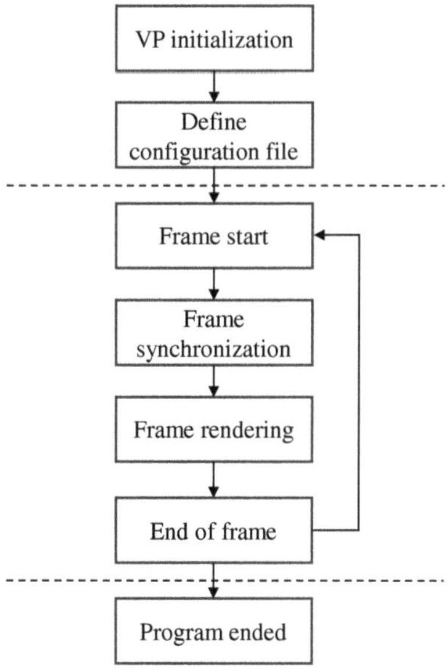

FIGURE 8.10 Execution workflow of the Vega Prime-based visual simulation program

(3) Thread Configuration: This completes the thread configuration in Vega Prime, establishing links among the involved classes.

(4) Simulation Loop: A frame loop is executed to refresh the visual scene display.

(5) Program Termination: The program concludes by properly shutting down Vega Prime, terminating active threads, and releasing allocated memory.

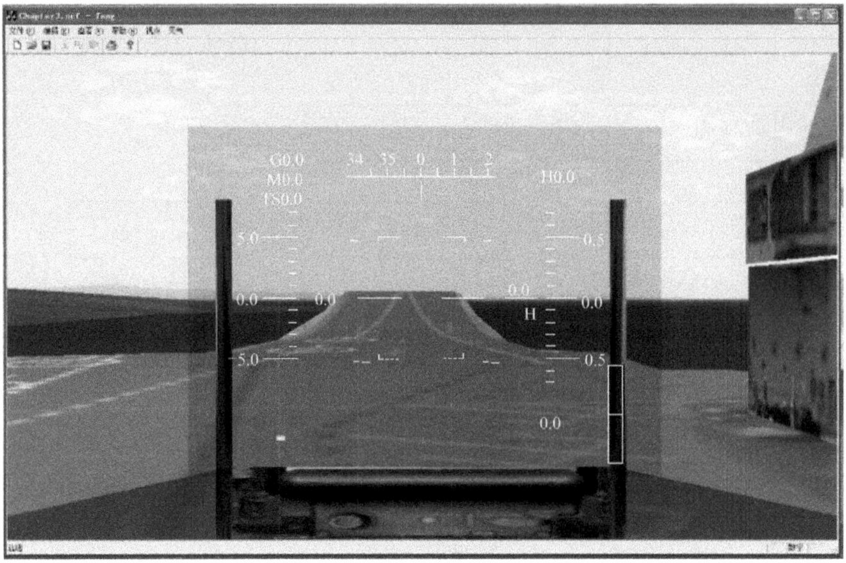

FIGURE 8.11 Visual simulation system interface

The interface of the visual simulation system is shown in Figure 8.11. A manned aircraft takes off from the runway on an aircraft carrier flight deck. The center of the interface simulates the aircraft's head-up display (HUD). Composed of white lines, numbers, and symbols, the HUD primarily presents flight-status and mission-parameter information.

8.4 HUMAN-MACHINE INTERACTION SYSTEM

The human-machine interaction (HMI) system serves as the channel for information exchange and control, enabling collaborative decision-making between manned aircraft and UAVs. The HMI system primarily consists of two subsystems, including the integrated situation awareness subsystem and the instrumentation subsystem.

8.4.1 Integrated Situation Awareness Subsystem

The integrated situation awareness subsystem is developed using Qt, a cross-platform C++ framework for graphical user interface applications. Due to its ease of use and extensive variety of interface controls, Qt is particularly suitable for simulation systems requiring intuitive visual feedback.

Figure 8.12 shows the simulation interface of the integrated situation awareness subsystem. The upper-left section of the interface displays the

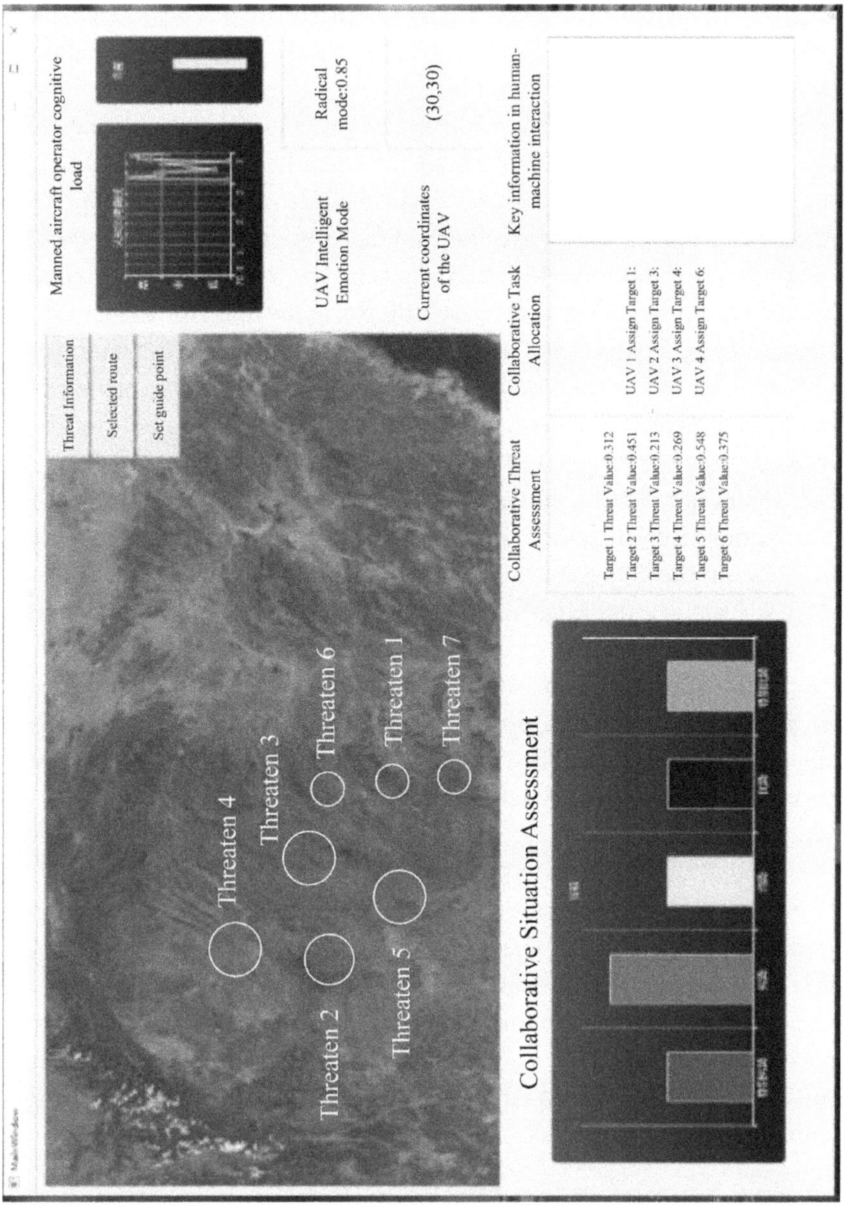

FIGURE 8.12 Integrated situation awareness subsystem simulation interface

battlefield situation for friendly forces, indicating the initial positions of friendly manned aircraft and UAVs. Circles represent the locations of known threats. The panels below present the simulation result interfaces for collaborative situation assessment, collaborative threat assessment, and collaborative mission assignment, respectively. The collaborative situation assessment simulation is displayed using five bar charts, from left to right, with different situation levels, including far disadvantage, disadvantage, parity, advantage, and far advantage for the friendly side. This visual presentation of situation assessment results helps the manned aircraft operator clearly capture the current friendly and enemy situations. In contrast, the results of collaborative threat assessment and collaborative mission assignment are provided in text format, allowing the operator to focus on and act upon them efficiently. Critical human-computer interaction information is displayed in the lower-right section of the interface, primarily including emergent threat alerts and decision intervention prompts for the manned aircraft operator.

The upper-right section of the interface displays the cognitive workload of the manned aircraft operator and the intelligent emotional model of the UAV. Clicking the "Operator Cognitive Workload" button on the interface reveals the eye-tracking monitoring data of the operator, as shown in Figure 8.13. The physiological (eye movement) data can be transmitted to the integrated situation-awareness subsystem via the TCP protocol using physiological monitoring equipment (eye tracker). This approach is adopted primarily as the TCP ensures both the integrity and real-time performance of the collected data transmission.

8.4.2 Instrumentation Subsystem

The instrumentation subsystem is developed using the Qt framework. The main development steps involve designing the corresponding instrument frames and loading instrument images to enhance their feasibility. By designing an appropriate driving program, the subsystem can accurately display data after receiving it. The instrumentation subsystem includes various indicators such as an airspeed indicator, attitude indicator, altimeter, and heading indicator, as shown in Figure 8.14.

8.5 SIMULATION EXPERIMENT VERIFICATION AND ANALYSIS

The simulation experiment procedure is as follows: the manned aircraft operator initially wears the eye tracker, which is set up by professionals to record real-time data. The pilot then operates one manned aircraft, leading

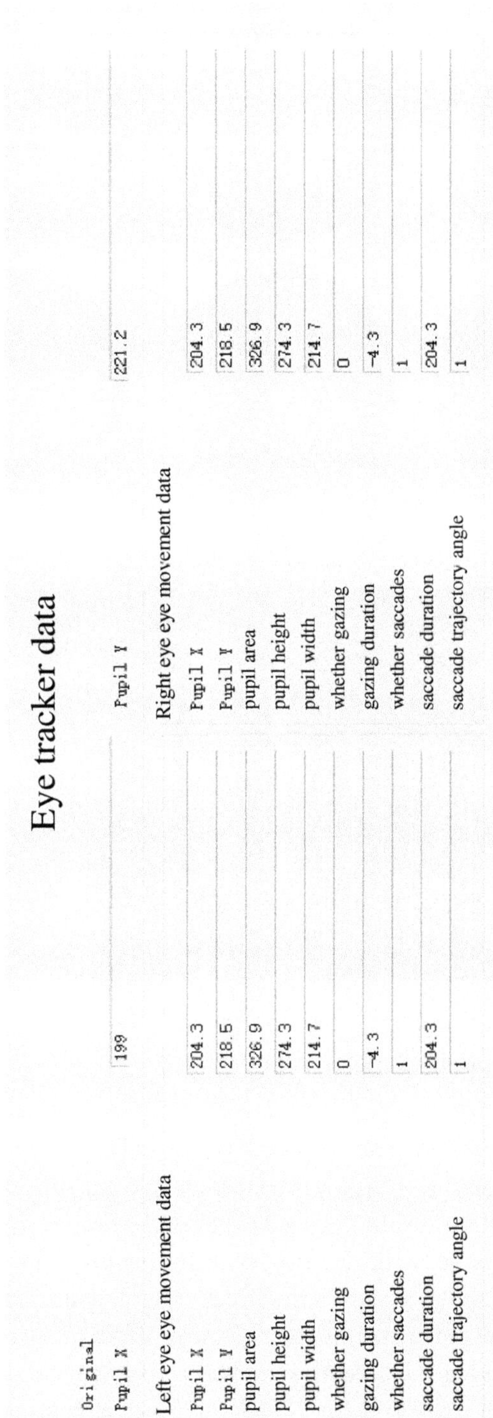

FIGURE 8.13 Eye-tracking data of the manned aircraft operator

FIGURE 8.14 Instrumentation subsystem

four UAVs to the mission area to execute the task. During the mission, the operator must monitor the display interfaces of various subsystems and, by perceiving the battlefield situation and engaging in human-machine interaction, execute collaborative decision-making between the manned aircraft and UAVs within a cognitive-intelligent interactive framework. After completion of the experiment, subjective rating scales are used to measure the operator's workload. Currently, the Subjective Workload Assessment Technique (SWAT) scale and the NASA-TLX assessment are commonly used due to their mature mechanism and high reliability.

8.5.1 The SWAT Scale

The SWAT scale comprises three dimensions, including time load (T), psychological stress load (S), and effort load (E), each rated on a three-level scale. Subjects rank these dimensions based on their experience during the experiment and select the appropriate state for each. Finally, based on the subject's ranking and selected levels, the corresponding sequence is located within the predefined grouping table to obtain the evaluation score. That is, the higher the score is, the greater the workload. The grouping and scoring criteria for the SWAT scale are presented in Table 8.5, and the SWAT questionnaire is shown in Figure 8.15.

TABLE 8.5 SWAT Scale Grouping and Scoring Criteria

Score	**Group**					
	TES	**TSE**	**ETS**	**EST**	**STE**	**SET**
1	1 1 1	1 1 1	1 1 1	1 1 1	1 1 1	1 1 1
2	1 1 2	1 2 1	1 1 2	2 1 1	1 2 1	2 1 1
3	1 1 3	1 3 1	1 1 3	3 1 1	1 3 1	3 1 1
4	1 2 1	1 1 2	2 1 1	1 1 2	2 1 1	1 2 1
5	1 2 2	1 2 2	2 1 2	2 1 2	2 2 1	2 2 1
6	1 2 3	1 3 2	2 1 3	3 1 2	2 3 1	3 2 1
7	1 3 1	1 1 3	3 1 1	1 1 3	3 1 1	1 3 1
8	1 3 2	1 2 3	3 1 2	2 1 3	3 2 1	2 3 1
9	1 3 3	1 3 3	3 1 3	3 1 3	3 3 1	3 3 1
10	2 1 1	2 1 1	1 2 1	1 2 1	1 1 2	1 1 2
11	2 1 2	2 2 1	1 2 2	2 2 1	1 2 2	2 1 2
12	2 1 3	2 3 1	1 2 3	3 2 1	1 3 2	3 1 2
13	2 2 1	2 1 2	2 2 1	1 2 2	2 1 2	1 2 2
14	2 2 2	2 2 2	2 2 2	2 2 2	2 2 2	2 2 2
15	2 2 3	2 3 2	2 2 3	3 2 2	2 3 2	3 2 2
16	2 3 1	2 1 3	3 2 1	1 2 3	3 1 2	1 3 2
17	2 3 2	2 2 3	3 2 2	2 2 3	3 2 2	2 3 2
18	2 3 3	2 3 3	3 2 3	3 2 3	3 3 2	3 3 2
19	3 1 1	3 1 1	1 3 1	1 3 1	1 1 3	1 1 3
20	3 1 2	3 2 1	1 3 2	2 3 1	1 2 3	2 1 3
21	3 1 3	3 3 1	1 3 3	3 3 1	1 3 3	3 1 3
22	3 2 1	3 1 2	2 3 1	1 3 2	2 1 3	1 2 3
23	3 2 2	3 2 2	2 3 2	2 3 2	2 2 3	2 2 3
24	3 2 3	3 3 2	2 3 3	3 3 2	2 3 3	3 2 3
25	3 3 1	3 1 3	3 3 1	1 3 3	3 1 3	1 3 3
26	3 3 2	3 2 3	3 3 2	2 3 3	3 2 3	2 3 3
27	3 3 3	3 3 3	3 3 3	3 3 3	3 3 3	3 3 3

SWAT Questionnaire

Name: ______ Group: ______

Select the level that best matches the current situation for each dimension below

I. Temporal Load (T)

1.Often have spare time. Interruptions or overlaps between activities rarely or never occur.

2.Occasionally have spare time. Interruptions or overlaps between activities occur frequently.

3.Almost no spare time. Interruptions or overlaps between activities are very frequent.

II. Effort Load (E)

1.Minimal conscious mental effort or focused attention required. Activities are almost automatic and require little attention.

2.Moderate conscious mental effort or concentration required. Activity complexity is moderately high due to uncertainty, unpredictability, or unfamiliarity, requiring considerable attention.

3.Extensive mental effort and concentration required. Highly complex activities demand full attention.

III. Psychological Stress Load (S)

1.Little confusion, risk, frustration, or anxiety exists, and adaptation is easily achieved.

2.Moderate stress due to confusion, frustration, or anxiety significantly increases workload, requiring substantial compensation to maintain adequate performance.

3.Very intense stress due to confusion, frustration, or anxiety. Requires strong determination and self-control.

Finally, rank the importance of these three dimensions:

① ______ ② ______ ③ ______

FIGURE 8.15 SWAT scale questionnaire

8.5.2 NASA-TLX Assessment

The NASA-TLX assessment is shown in Figure 8.16 and consists of the following six factors: mental demand, physical demand, temporal demand, performance, effort, and frustration. Scoring may occur during a task, after individual tasks, or upon completion of the entire experiment. Each scale is divided into 20 equal intervals, ranging from "very low" (0) to "very high" (100), with increments of 5 per interval.

Six postgraduate students from relevant disciplines participated in the project as subjects (serving as manned aircraft operators). Using the two aforementioned scales, they assessed their own cognitive status after performing collaborative decision-making tasks under both conventional

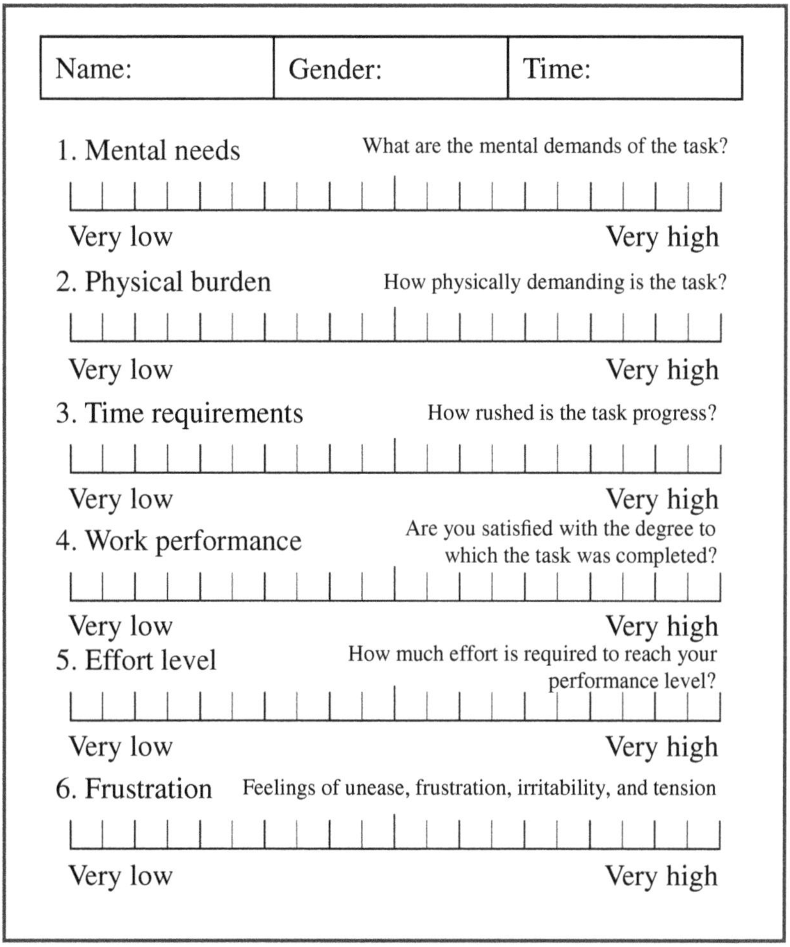

Name:	Gender:	Time:

1. Mental needs — What are the mental demands of the task?
Very low — Very high

2. Physical burden — How physically demanding is the task?
Very low — Very high

3. Time requirements — How rushed is the task progress?
Very low — Very high

4. Work performance — Are you satisfied with the degree to which the task was completed?
Very low — Very high

5. Effort level — How much effort is required to reach your performance level?
Very low — Very high

6. Frustration — Feelings of unease, frustration, irritability, and tension
Very low — Very high

FIGURE 8.16 NASA-TLX assessment questionnaire

interaction methods and the intelligent interaction approach. The results are presented in Table 8.6.

Based on the analysis of Table 8.6, the participants' workload scores were plotted, as shown in Figure 8.17. It can be observed that, under the cognitive intelligent interaction mode, the performance scores of the subjects are generally higher than those under the conventional interaction mode. Meanwhile, the cognitive workload of the manned aircraft operators is generally lower than that in the conventional mode, which contributes to enhanced task efficacy in collaborative decision-making between manned aircraft and UAVs.

TABLE 8.6 Comparison of Manned Aircraft Operator Evaluation

No. of Subject	Time	Cognitive Intelligent Interaction System Score	Score of Ordinary Interactive System
1	Morning	505	473
	Noon	528	482
	Night	486	458
	Morning	558	543
2	Noon	569	530
	Night	538	507
	Morning	542	490
3	Noon	537	520
	Night	490	478
	Morning	478	464
4	Noon	530	513
	Night	486	459
	Morning	524	473
5	Noon	492	497
	Night	479	466
	Morning	528	510
6	Noon	490	484
	Night	526	515

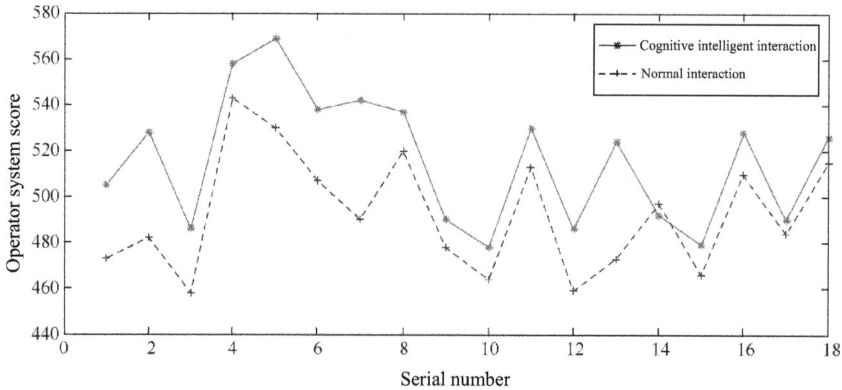

FIGURE 8.17 Workload scores of participants

8.6 CHAPTER SUMMARY

The main works completed in this chapter include:

(1) Designed a construction scheme for the cognitive intelligent interactive simulation system of manned aircraft and UAV collaborative decision-making.

(2) Introduced the hardware and software implementation methods of the flight control system, visual system, and human-machine interaction system.

(3) Conducted human-in-the-loop simulation experiments for manned aircraft and UAV cooperative decision-making tasks and performed data analysis using the NASA-TLX assessment and the subjective workload assessment technique scale.

8.6 CHAPTER SUMMARY

References

[1] Huang, J. Z., Zhou, D. Y., & Zeng, L. N. (2015). Research Status and Implications of Manned and Unmanned Aircraft Cooperative Combat Abroad. *Proceedings of the 4th China Airborne Weapon Conference*, 1–4.

[2] United States Department of Defense. (2007). *Unmanned Systems Roadmap 2007–2032*. https://rosap.ntl.bts.gov/view/dot/18247

[3] Zhu, H. Y., Zhang, Q. J., & Shen, L. C. (2010). Technologies for Promoting Cooperatively Combat Capability of UCAV. *Control Engineering of China*, 17(S1), 52–55. https://doi.org/10.14107/j.cnki.kzgc.2010.s1.005

[4] Qiu, X. J., Chen, J., & Zhang, X. W. (2015). Supervisory Mission Control Method for Unmanned Vehicle Based on RBFCM. *Proceedings of the 4th China Airborne Weapon Conference*, 1–5.

[5] Zhao, C. L., Zhu, R. G., & Chen, S. D. (2005). The Study of Manned Fighter/UCAV Cooperative Attacking System. *Proceedings of the 2nd Sino-Russian UAV Academic Conference*.

[6] Chen, Z. J., Wei, J. Z., Wang, Y. X., & Zhou, R. (2011). UAV Autonomous Control Levels and System Structure. *Acta Aeronautica et Astronautica Sinica*, 32(6), 1075–1083.

[7] Redding, J., Boskovic, J., Mehra, R., & Rui, C. (2008). Heterogeneous Cooperative Control of Multiple UAVs with Collaborative Assignment and Reactive Motion Planning. *AIAA Guidance, Navigation and Control Conference and Exhibit*, 6794. https://doi.org/10.2514/6.2008-6794

[8] Niu, Y. F., Shen, L. C., Li, J., & Wang, X. (2019). Key Scientific Problems in Cooperation Control of Unmanned-Manned Aircraft Systems. *Sci Sin Inform*, 49(5), 538–554. https://doi.org/10.1360/N112019-00008

[9] Zhang, X. D., Sun, Z. W., Wu, L., Hao, M., & Yin, H. (2021). Research on the Fundamental and Grand Concept of Intelligent Cooperative Combat of Manned/Unmanned Aircraft. *Unmanned Systems Technology*, 4(2), 62–68. https://doi.org/10.19942/j.issn.2096-5915.2021.2.020

[10] Gu, H. Y., & Xu, C. (2017). Cooperative Operational Technologies for Manned/Unmanned Teaming. *Command Information System and Technology*, 8(6), 33–41. https://doi.org/10.15908/j.cnki.cist.2017.06.006

[11] Liu, L. J., & Tu, T. J. (2018). Analysis of Typical Cooperative Combat Patterns of Shipborne Manned Helicopter/UAV. *Electronic Technology & Software Engineering*, (13), 96–98.

[12] Sarkesain, J. F., & O'Brien, T. W. (2014). A Framework for Achieving Dynamic Cyber Effects Through Distributed Cyber Command and Control/Battle Management (C2/BM). *Modeling and Simulation Support for System of Systems Engineering Applications* (pp. 531–564). https://doi.org/10.1002/9781118501757.ch20

[13] Li, L. (2020). Development Analysis of Typical Manned/Unmanned Aerial Vehicle Collaborative Operations Projects Abroad. *Unmanned Systems Technology*, 3(4), 83–90. https://doi.org/10.19942/j.issn.2096-5915.2020.04.010

[14] Shen, C., Li, L., Wu, Y., & Liu, D. (2018). Research on the Capability of the U.S. Manned/Unmanned Autonomous Collaborative Operations. *Tactical Missile Technology*, (1), 16–21. https://doi.org/10.16358/j.issn.1009-1300.2018.01.03

[15] Song, C., & Zhang, P. P. (2019). Characteristics and Impacts of Manned and Unmanned Cooperative Air Combat. *Aeronautical Missile*, 420(12), 78–81.

[16] Valenti, M., Schouwenaars, T., Kuwata, Y., Feron, E., How, J., & Paunicka, J. (2004). Implementation of a Manned Vehicle-UAV Mission System. *AIAA Guidance, Navigation, and Control Conference and Exhibit*, 5142. https://doi.org/10.2514/6.2004-5142

[17] Schouwenaars, T., Valenti, M., Feron, E., How, J., & Roche, E. (2006). Linear Programming and Language Processing for Human-Unmanned Aerial-Vehicle Team Missions. *Journal of Guidance, Control, and Dynamics*, 29(2), 303–313. https://doi.org/10.2514/1.13162

[18] Defense Advanced Research Projects Agency. (n.d.). *Broad Agency Announcement: System of Systems Integration Technology and Experimentation*. www.darpa.mil/program/system-of-systems-integration-technology-and-experimentation

[19] Lummus, R. (2003). Mission Battle Management System Fighter Engagement Manager Concept. *AIAA International Air and Space Symposium and Exposition: The Next 100 Years*, 2857. https://doi.org/10.2514/6.2003-2857

[20] Endsley, M. (2015). *Autonomous Horizons: System Autonomy in the Air Force – A Path to the Future (Volume I: Human Autonomy Teaming)*. U.S. Department of the Air Force. https://doi.org/10.13140/RG.2.1.1164.2003

[21] Chen, J., & Xin, B. (2018). Key Scientific Problems in the Autonomous Cooperation of Manned-Unmanned Systems. *Sci Sin Inform*, 48, 1270–1274. https://doi.org/10.1360/N112018-00092

[22] Lockheed Martin. (2017). *U.S. Air Force, Lockheed Martin Demonstrate Manned/Unmanned Teaming*. https://news.lockheedmartin.com/2017-04-10-U-S-Air-Force-Lockheed-Martin-Manned-Unmanned-Teaming

[23] Han, W., Wu, L. Y., & Zhang, Y. (2019). Review and Prospect of Flight Control of Carrier-Based Aircraft/Unmanned Aerial Vehicles Formation. *Science Technology and Engineering*, 19(36), 73–80.

[24] United States Department of Defense. (2017). *Unmanned Systems Integrated Roadmap: Fiscal Years 2017–2042*. https://digital.library.unt.edu/ark:/67531/metadc1982990/

[25] Xu, J. (2017). *Autonomous Ground Attack Decision-Making of UCAV Based on FCM Model Theory* [Master thesis, Northwestern Polytechnical University].

[26] Zhang, G. Z., Shen, L. C., & Zhu, H. Y. (2009). The Current Situation and Enlightenment of Supervisory Control Technology for Multiple UAVs. *National Defense Technology*, 30(4), 5–10. https://doi.org/10.3969/j.issn.1671-4547.2009.04.002

[27] Duan, H. B., Zhang, D. F., Fan, Y. M., & Deng, Y. (2019). From Wolf Pack Intelligence to UAV Swarm Cooperative Decision-Making. *Sci Sin Inform*, 49(1), 112–118. https://doi.org/10.1360/N112018-00168

[28] Endsley, M. R. (2015). Situation Awareness Misconceptions and Misunderstandings. *Journal of Cognitive Engineering and Decision Making*, 9(1), 4–32. https://doi.org/10.1177/1555343415572631

[29] Kosko, B. (1986). Fuzzy Cognitive Maps. *International Journal of Man-Machine Studies*, 24(1), 65–75. https://doi.org/10.1016/S0020-7373(86)80040-2

[30] Amirkhani, A., Papageorgiou, E. I., Mohseni, A., & Mosavi, M. R. (2017). A Review of Fuzzy Cognitive Maps in Medicine: Taxonomy, Methods, and Applications. *Computer Methods and Programs in Biomedicine*, 142, 129–145. https://doi.org/10.1016/j.cmpb.2017.02.021

[31] Zdanowicz, P., & Petrovic, D. (2017). New Mechanisms for Reasoning and Impacts Accumulation for Rule-Based Fuzzy Cognitive Maps. *IEEE Transactions on Fuzzy Systems*, 26(2), 543–555. https://doi.org/10.1109/TFUZZ.2017.2686363

[32] Harmati, I. Á., & Kóczy, L. T. (2020). On the Convergence of Fuzzy Grey Cognitive Maps. In P. Kulczycki, J. Kacprzyk, L. Kóczy, R. Mesiar, & R. Wisniewski (Eds.), *Information Technology, Systems Research, and Computational Physics*. Springer. https://doi.org/10.1007/978-3-030-18058-4_6

[33] Hajek, P., & Froelich, W. (2019). Integrating TOPSIS with Interval-Valued Intuitionistic Fuzzy Cognitive Maps for Effective Group Decision Making. *Information Sciences*, 485, 394–412. https://doi.org/10.1016/j.ins.2019.02.035

[34] Arruda, L. V. R., Mendonça, M., Neves, F., Chrun, I. R., & Papageorgiou, E. I. (2016). Artificial Life Environment Modeled by Dynamic Fuzzy Cognitive Maps. *IEEE Transactions on Cognitive and Developmental Systems*, 10(1), 88–101. https://doi.org/10.1109/TCDS.2016.2634865

[35] Bourgani, E., Manis, G., Stylios, C. D., & Georgopoulos, V. C. (2016). Timed-Fuzzy Cognitive Maps: An Overview. *2016 IEEE International Conference on Systems, Man, and Cybernetics (SMC)*, 4483–4488. https://doi.org/10.1109/SMC.2016.7844938

[36] Papageorgiou, K. I., Poczeta, K., Papageorgiou, E., Gerogiannis, V. C., & Stamoulis, G. (2019). Exploring an Ensemble of Methods That Combines Fuzzy Cognitive Maps and Neural Networks in Solving the Time Series Prediction Problem of Gas Consumption in Greece. *Algorithms*, 12(11), 235. https://doi.org/10.3390/a12110235

[37] Felix, G., Nápoles, G., Falcon, R., Froelich, W., Vanhoof, K., & Bello, R. (2019). A Review on Methods and Software for Fuzzy Cognitive Maps. *Artificial Intelligence Review*, 52, 1707–1737. https://doi.org/10.1007/s10462-017-9575-1

[38] Khodadadi, M., Shayanfar, H., Maghooli, K., & Hooshang Mazinan, A. (2019). Fuzzy Cognitive Map Based Approach for Determining the Risk of Ischemic Stroke. *IET Systems Biology*, 13(6), 297–304. https://doi.org/10.1049/iet-syb.2018.5128

[39] Yang, Z., & Liu, J. (2020). Learning Fuzzy Cognitive Maps with Convergence Using a Multi-Agent Genetic Algorithm. *Soft Computing*, 24(6), 4223–4234. https://doi.org/10.1007/s00500-019-04173-2

[40] Chi, Y., & Liu, J. (2019). Reconstructing Gene Regulatory Networks with a Memetic-Neural Hybrid Based on Fuzzy Cognitive Maps. *Natural Computing*, 18, 301–312. https://doi.org/10.1007/s11047-016-9547-4

[41] Bahri, O., Mourhir, A., & Papageorgiou, E. I. (2020). Integrating Fuzzy Cognitive Maps and Multi-Agent Systems for Sustainable Agriculture. *Euro-Mediterranean Journal for Environmental Integration*, 5, 7. https://doi.org/10.1007/s41207-020-0143-8

[42] Naskali, Y. K., Gürbüz, T., & Albayrak, Y. (2019). Human Reliability Assessment in Healthcare Operations Using Fuzzy Cognitive Maps. *Journal of Multiple-Valued Logic & Soft Computing*, 32.

[43] Chen, J., Xu, J., & Gao, X. G. (2017). Autonomous Attack Decision-Making of UCAV Based on ABFCM Model Framework. *Systems Engineering and Electronics*, 39(3), 549–556.

[44] Zieba, S., & Polet, P. (2011). Using Adjustable Autonomy and Human-Machine Cooperation to Make a Human-Machine System Resilient-Application to a Ground Robotic System. *Information Sciences*, 181(3), 379–397. https://doi.org/10.1016/j.ins.2010.09.035

[45] Cummings, M. L., & Guerlain, S. (2004). Human Performance Issues in Supervisory Control of Autonomous Airborne Vehicles. *Proceedings of AUVSI Unmanned Systems North America Conference*, Anaheim.

[46] Bertuccelli, L. F., & Cummings, M. L. (2012). Operator Choice Modeling for Collaborative UAV Visual Search Tasks. *IEEE Transactions on Systems, Man, and Cybernetics – Part A: Systems and Humans*, 42(5), 1088–1099. https://doi.org/10.1109/TSMCA.2012.2189875

[47] Rudnick, G., Claus, S., & Schulte, A. (2014). Flight Testing of Agent Supervisory Control on Heterogeneous Unmanned Aerial System Platforms. *2014 IEEE/AIAA 33rd Digital Avionics Systems Conference (DASC)*, Colorado Springs. https://doi.org/10.1109/DASC.2014.6979468.

[48] Schmitt, F., & Schulte, A. (2015). Mixed-Initiative Mission Planning Using Planning Strategy Models in Military Manned-Unmanned Teaming Missions. *Proceedings of IEEE International Conference on Systems, Man, and Cybernetics*, Hong Kong. https://doi.org/10.1109/SMC.2015.247.

[49] Bevacqua, G., & Cacace, J. (2015). Mixed-Initiative Planning and Execution for Multiple Drones in Search and Rescue Missions. *Proceedings of the 25th International Conference on Automated Planning and Scheduling*, Jerusalem. https://doi.org/10.1609/icaps.v25i1.13700

[50] Dorton, S., Terry, B., Jaeger, B., & Shearer, P. B. (2016). Development of a Recognition Primed Decision Agent for Supervisory Control of Autonomy. *2016 IEEE International Multi-Disciplinary Conference on Cognitive Methods in Situation Awareness and Decision Support (CogSIMA)*, San Diego. https://doi.org/10.1109/COGSIMA.2016.7497806.

[51] Lim, H. J., Choi, S. H., Oh, J., Kim, S., Lee, J., Kim, B. S., Kim, S., & Yang, J. H. (2017). Comparison Study of Potential Workload Index in A Simulated Multiple-UAV Operation Environment. *2017 17th International Conference on Control*, Jeju. https://doi.org/10.23919/ICCAS.2017.8204405

[52] Cabezas, R. T., & Verdegay, J. L. (2019). Intelligent System of Strategic Monitoring. *14th Iberian Conference on Information Systems and Technologies (CISTI)*, Coimbra. https://doi.org/10.23919/CISTI.2019.8760643

[53] Yin, H., Hou, T. T., & Li, D. G. (2021). Cooperative Effectiveness Evaluation of MAV/UAV for Typical Missions. *Unmanned Systems Technology*, 4(1), 62–70. https://doi.org/10.19942/j.issn.2096-5915.2021.1.008

[54] Chen, J., Zhang, X. W., Xu, J., & Gao, X. G. (2015). Human/Unmanned-Aerial-Vehicle Team Collaborative Decision-Making with Limited Intervention. *Acta Aeronautica et Astronautica Sinica*, 36(11), 3652–3665.

[55] Zhong, Y., Yao, P. Y., Wan, L. J., & Yang, J. (2018). Intervention Decision-Making in MAV/UAV Cooperative Engagement Based on Human Factors Engineering. *Journal of Systems Engineering and Electronics*, 29(3), 530–538. https://doi.org/10.21629/JSEE.2018.03.10

[56] Rogers, A., Ramchurn, S. D., & Jennings, N. R. (2012). Delivering the Smart Grid: Challenges for Autonomous Agents and Multi-Agent Systems Research. *AAAI Conference on Artificial Intelligence*, Toronto. https://doi.org/10.1609/aaai.v26i1.8445

[57] Yin, L., Zhang, R., & Gu, H. (2012). Situation Reasoning for an Adjustable Autonomy System. *International Journal of Intelligent Computing & Cybernetics*, 5(2), 226–238. https://doi.org/10.1108/17563781211231552

[58] Dumond, D., Ayers, J., Schurr, N., Carlin, A., Burke, D., & Rousseau, J. (2012). Coordinating with Humans by Adjustable-Autonomy for Multirobot Pursuit (CHAMP). *International Society for Optics and Photonics*, Baltimore. https://doi.org/10.1117/12.920451

[59] Brooks, R. (1985). A Robust Layered Control System for A Mobile Robot. *IEEE Journal on Robotics & Automation*, 2(1), 14–23. https://doi.org/10.1109/JRA.1986.1087032

[60] Duan, L. Q., Feng, P. M., & Zhang, R. B. (2011). Research on the Resilience of Human-machine System Based on Adjustable Autonomy. *Journal of Huazhong University of Science and Technology (Natural Science Edition)*, 39(S2), 72–76. https://doi.org/10.13245/j.hust.2011.s2.022

[61] Gadallah, A. M., & Hefny, H. A. (2010). Fuzzy Cognitive Map with Dynamic Fuzzification and Causality Behaviors. *2010 The 7th International Conference on Informatics and Systems (INFOS)*, Cairo.

[62] Tsadiras, A. K., & Margaritis, K. G. (1997). Cognitive Mapping and Certainty Neuron Fuzzy Cognitive Maps. *Information Sciences*, 101(97), 109–130. https://doi.org/10.1016/S0020-0255(97)00001-7

[63] Miao, C. Y., Goh, A., Miao, Y., & Yang, Z. H. (2002). Agent That Models, Reasons and Makes Decisions. *Knowledge-Based Systems*, 15(3), 203–211. https://doi.org/10.1016/S0950-7051(01)00157-5

[64] Levulis, S. J., Kim, S. Y., & DeLucia, P. R. (2016). Effects of Touch, Voice, and Multimodal Input on Multiple-UAV Monitoring During Simulated Manned-Unmanned Teaming in a Military Helicopter. *Proceedings of the Human Factors and Ergonomics Society Annual Meeting*, Washington, DC. https://doi.org/10.1177/0018720818788995

[65] Lim, Y., Gardi, A., Sabatini, R., Ramasamy, S., Kistan, T., Ezer, N., Vince, J., & Bolia, R. (2018). Avionics Human-Machine Interfaces and Interactions for Manned and Unmanned Aircraft. *Progress in Aerospace Sciences*, 91, 755–774. https://doi.org/10.1016/j.paerosci.2018.05.002

[66] Li, D. Y., Liu, M., Zhao, F., & Liu, Y. (2019). Challenges and Countermeasures of Interaction in Autonomous Vehicles. *Science China Information Sciences*, 62(5), 50201. https://doi.org/10.1007/s11432-018-9766-3

[67] Pacaux-Lemoine, M. P., Trentesaux, D., Rey, G. Z., & Millot, P. (2017). Designing Intelligent Manufacturing Systems through Human-Machine Cooperation Principles: A Human-Centered Approach. *Computers & Industrial Engineering*, 111, 581–595. https://doi.org/10.1016/j.cie.2017.05.014

[68] Dul, J., Bruder, R., Buckle, P., Carayon, P., Falzon, P., Marras, W. S., Wilson, J. R., & van der Doelen, B. (2012). A Strategy for Human Factors/Ergonomics: Developing the Discipline and Profession. *Ergonomics*, 55(4), 377–395. https://doi.org/10.1080/00140139.2012.661087

[69] Dul, J., & Neumann, W. P. (2009). Ergonomics Contributions to Company Strategies. *Applied Ergonomics*, 40(4), 745–752. https://doi.org/10.1016/j.apergo.2008.07.001

[70] Enderwick, T. P., & Meister, D. (2001). *Human Factors in System Design, Development, and Testing*. Boca Raton: CRC Press. https://doi.org/10.1201/b12452

[71] Stanton, N. A., Salmon, P. M., & Rafferty, L. A. (2017). *Human Factors Methods: A Practical Guide for Engineering and Design*. Boca Raton: CRC Press. https://doi.org/10.1201/9781315587394

[72] Karwowski, W. (2005). Ergonomics and Human Factors: The Paradigms for Science, Engineering, Design, Technology and Management of Human-Compatible Systems. *Ergonomics*, 48(5), 436–463. https://doi.org/10.1080/00140130400029167

[73] Moray, N. (1982). Subjective Mental Workload. *Human Factors*, 24(1), 25–40.

[74] Waard, D. D. (1996). *The Measurement of Drivers' Mental Workload*. Netherlands: Groningen University, Traffic Research Center.

[75] Chandler, P., & Sweller, J. (1991). Cognitive Load Theory and the Format of Instruction. *Cognition and Instruction*, 8(4), 293–332. https://doi.org/10.1207/s1532690xci0804_2

[76] Phillips, J. L., Shiffrin, R. M., & Atkinson, R. C. (1967). Effects of List Length on Short-term Memory. *Journal of Verbal Learning and Verbal Behavior*, 6(3), 303–311. https://doi.org/10.1016/S0022-5371(67)80117-8

[77] Ren, T. Z., Zhou, R., & Li, H. (2017). An Emotional Intelligence Based Autonomous Decision-Making Method for UAVs. *Electronics Optics & Control*, 24(3), 11–15, 19. https://doi.org/10.3969/j.issn.1671-637x.2017.03.003

[78] Qu, G., Dong, Y. F., & Yue, Y. (2016). Operational Effectiveness Evaluation of Ground Attack UCAV. *Fire Control & Command Control*, 41(4), 145–149.

[79] Wang, Y., Xiao, Y., Zhou, Q. X., Zhang, L. W., Tang, X. Z., & Zhang, H. (2018). Research Progress of Mental Workload Monitoring Technology Based on EEG. *Space Medicine & Medical Engineering*, 31(5), 577–582. https://doi.org/10.16289/j.cnki.1002-0837.2018.05.015

[80] Hayton, J. C., Allen, D. G., & Scarpello, V. (2004). Factor Retention Decisions in Exploratory Factor Analysis: A Tutorial on Parallel Analysis. *Organizational Research Methods*, 7(2), 191–205. https://doi.org/10.1177/1094428104263675

[81] Haapalainen, E., Kim, S. J., Forlizzi, J. F., & Dey, A. K. (2010). Psycho-Physiological Measures for Assessing Cognitive Load. *Proceedings of the 12th ACM International Conference on Ubiquitous Computing*, Copenhagen. https://doi.org/10.1145/1864349.186439

[82] Zarjam, P., Epps, J., & Lovell, N. H. (2015). Beyond Subjective Self-Rating: EEG Signal Classification of Cognitive Workload. *IEEE Transactions on Autonomous Mental Development*, 7(4), 301–310. https://doi.org/10.1109/TAMD.2015.2441960

[83] Wang, S., Gwizdka, J., & Chaovalitwongse, W. A. (2015). Using Wireless EEG Signals to Assess Memory Workload in The N-Back Task. *IEEE Transactions on Human-Machine Systems*, 46(3), 424–435. https://doi.org/10.1109/THMS.2015.2476818

[84] Almahasneh, H., Chooi, W. T., Kamel, N., & Malik, A. S. (2014). Deep in Thought while Driving: An EEG Study on Drivers' Cognitive Distraction. *Transportation Research Part F: Traffic Psychology and Behaviour*, 26, 218–226. https://doi.org/10.1016/j.trf.2014.08.001

[85] Chen, S., & Epps, J. (2013). Automatic Classification of Eye Activity for Cognitive Load Measurement with Emotion Interference. *Computer Methods and Programs in Biomedicine*, 110(2), 111–124. https://doi.org/10.1016/j.cmpb.2012.10.021

[86] Palinko, O., Kun, A. L., Shyrokov, A., & Heeman, P. (2010). Estimating Cognitive Load Using Remote Eye Tracking in a Driving Simulator. *Proceedings of the 2010 Symposium on Eye-tracking Research & Applications*, Austin. https://doi.org/10.1145/1743666.1743701

[87] Rosch, J. L., & Vogel-Walcutt, J. J. (2013). A Review of Eye-Tracking Applications as Tools for Training. *Cognition, Technology & Work*, 15(3):,313–327. https://doi.org/10.1007/s10111-012-0234-7

[88] Ma, R. R. (2010). *Research of Ensemble Learning*. Shandong University of Science and Technology.

[89] Cui, L. J., Li, K., & Ni, Z. H. (2007). A Study of Ensemble Learning Algorithm for Classification. *Journal of Hebei University (Natural Science Edition)*, 2007(4), 423–427. https://doi.org/10.3969/j.issn.1000-1565.2007.04.020

[90] Li, G. Z., & Li, D. (2007). Feature Selection for Ensemble Learning. *Journal of Shanghai University (Natural Science Edition)*, 2007(5), 598–604.

[91] Zhong, Y., Zhang, J., & Deng, C. (2017). Cooperative Engagement Problems About Manned/Unmanned Aerial Vehicles. *Command Information System and Technology*, 8(4), 19–25. https://doi.org/10.15908/j.cnki.cist.2017.04.003

[92] Hu, J., Huang, C., Zhao, H., & Zang, X. (2010). On UCAV's Situation Assessment Method Based on Variable Precision Rough Set Theory. *Electronics Optics & Control*, 17(3), 23–26, 43.

[93] Zhao, K., Huang, C., Wei, Z., & Wang, L. (2019). Situation Assessment for Unmanned Aerial Vehicle Air Combat Based on Anti-Reasoning Rules Decision Tree. *Journal of Harbin Institute of Technology*, 51(4), 66–73. https://doi.org/10.11918/j.issn.0367-6234.201801006

[94] Li, C., Duanmu, J., Lei, Y., & Gao, J. (2012). Situation Assessment Based on Cognitive Maps and Intuitionistic Fuzzy Reasoning. *Systems Engineering and Electronics*, 34(10), 2064–2068. https://doi.org/10.3969/j.issn.1001-506X.2012.10.15

[95] Fu, T., & Wang, J. (2016). Threat Assessment of Aerial Targets in Air-Defense. *Command Control & Simulation*, 38(3), 63–69. https://doi.org/10.3969/j.issn.1673-3819.2016.03.012

[96] Liu, H. Q., Wei, X. Z., Fu, Z. W., & Zhou, Z. L. (2013). Cooperative Task Assignment of Manned/Unmanned Aerial Vehicle Formation in Air Combat. *Electronics Optics & Control*, 20(6),16–19.

[97] Wang, K. (2009). *Research on Human-Machine Dynamic Function Allocation of Multi-UAV Supervisory Control System* [Master thesis, National University of Defense Technology].

[98] Di, B., Zhou, R., & Ding, Q. (2013). Distributed Coordinated Heterogeneous Task Allocation for Unmanned Aerial Vehicles. *Control and Decision*, 28(2), 274–278. https://doi.org/10.13195/j.cd.2013.02.117.dib.010

[99] Wan, L., Yao, P., & Sun, P. (2013). Distributed Task Allocation Method of Manned/Unmanned Combat Agents. *Systems Engineering and Electronics*, 35(2), 310–316. https://doi.org/10.3969/j.issn.1001-506X.2013.02.13

[100] Ma, X., Chen, X., & Lei, Y. (2009). The Data Link Based A* Algorithm Used for UCAV Path Planning. *Electronics Optics & Control*, 16(12), 15–17, 21.

[101] Luo, W., & Wei, R. (2019). Intelligent Route Planning for Cooperative Striking of Manned/Unmanned Aerial Vehicle. *Control Theory & Applications*, 36(7), 1090–1095.

[102] Wang, W., Wang, W., & Hou, X. (2017). *Vega Prime Development and Simulation Application*. Southwest Jiaotong University Press.

www.ingramcontent.com/pod-product-compliance
Ingram Content Group UK Ltd.
Pitfield, Milton Keynes, MK11 3LW, UK
UKHW022223190926
475702UK00003B/478
9781041341796